GIS Algorithms

Theory and Applications for Geographic Information Science & Technology

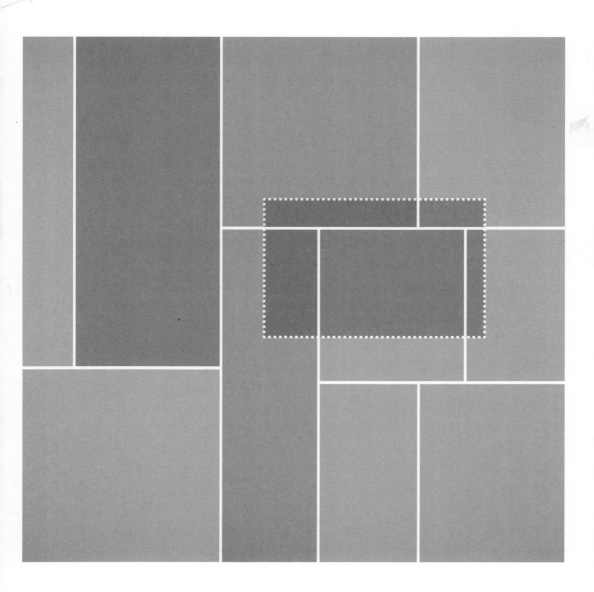

Geographic Information Science and Technology (GIST) is enjoying profound innovation in its research and development. *SAGE Advances in GIST* will provide students with learning resources to support and enrich their interest in the workings of Geographic Information Science and Technology. These highly visual and highly applied texts will promote curiosity about developments in the field, as well as provide focused and up-to-date tools for doing GIS research.

Series edited by Mei-Po Kwan, Department of Geography, University of California, Berkeley

Also in the GIST series:

Spatial Statistics and Geostatistics
Yongwan Chun and Daniel A. Griffith

GIS Algorithms

Ningchuan Xiao

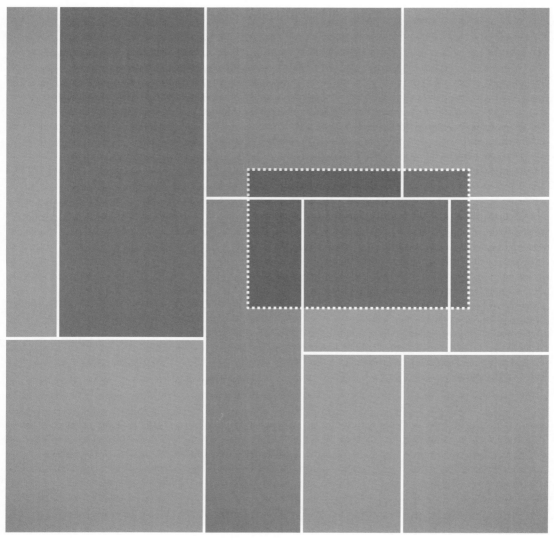

Los Angeles | London | New Delhi
Singapore | Washington DC

Los Angeles | London | New Delhi
Singapore | Washington DC

SAGE Publications Ltd
1 Oliver's Yard
55 City Road
London EC1Y 1SP

SAGE Publications Inc.
2455 Teller Road
Thousand Oaks, California 91320

SAGE Publications India Pvt Ltd
B 1/I 1 Mohan Cooperative Industrial Area
Mathura Road
New Delhi 110 044

SAGE Publications Asia-Pacific Pte Ltd
3 Church Street
#10-04 Samsung Hub
Singapore 049483

Editor: Robert Rojek
Editorial assistant : Matt Oldfield
Production editor: Katherine Haw
Copyeditor: Richard Leigh
Proofreader: Richard Hutchinson
Indexer: Bill Farrington
Marketing manager: Michael Ainsley
Cover design: Francis Kenney
Typeset by: C&M Digitals (P) Ltd, Chennai, India
Printed and bound by CPI Group (UK) Ltd,
   Croydon, CR0 4YY

© Ningchuan Xiao 2016

First published 2016

Apart from any fair dealing for the purposes of research or private study, or criticism or review, as permitted under the Copyright, Designs and Patents Act, 1988, this publication may be reproduced, stored or transmitted in any form, or by any means, only with the prior permission in writing of the publishers, or in the case of reprographic reproduction, in accordance with the terms of licences issued by the Copyright Licensing Agency. Enquiries concerning reproduction outside those terms should be sent to the publishers.

Library of Congress Control Number: 2015940434

British Library Cataloguing in Publication data

A catalogue record for this book is available from the British Library

ISBN 978-1-4462-7432-3
ISBN 978-1-4462-7433-0 (pbk)

At SAGE we take sustainability seriously. Most of our products are printed in the UK using FSC papers and boards. When we print overseas we ensure sustainable papers are used as measured by the PREPS grading system. We undertake an annual audit to monitor our sustainability.

To Alester, for understanding everything.

Contents

About the Author   x
Preface   xi

1 Introduction   1
    1.1 Computational concerns for algorithms   2
    1.2 Coding   6
    1.3 How to use this book   7

I GEOMETRIC ALGORITHMS   9

2 Basic Geometric Operations   11
    2.1 Point   11
    2.2 Distance between two points   13
    2.3 Distance from a point to a line   15
    2.4 Polygon centroid and area   18
    2.5 Determining the position of a point with respect to a line   20
    2.6 Intersection of two line segments   22
    2.7 Point-in-polygon operation   26
       2.7.1 Even–odd algorithm   27
       2.7.2 Winding number algorithm   30
    2.8 Map projections   33
    2.9 Notes   46
    2.10 Exercises   47

3 Polygon Overlay   49
    3.1 Line segment intersections   49
    3.2 Overlay   58
    3.3 Notes   66
    3.4 Exercises   67

II SPATIAL INDEXING   69

4 Indexing   71
    4.1 Exercises   76

5 *k*-D Trees — 77
- 5.1 Point *k*-D trees — 77
  - 5.1.1 Orthogonal range query — 82
  - 5.1.2 Circular range query — 84
  - 5.1.3 Nearest neighbor query — 85
- 5.2 Point region *k*-D trees — 87
- 5.3 Testing *k*-D trees — 93
- 5.4 Notes — 97
- 5.5 Exercises — 98

6 Quadtrees — 99
- 6.1 Region quadtrees — 99
- 6.2 Point quadtrees — 105
- 6.3 Notes — 110
- 6.4 Exercises — 111

7 Indexing Lines and Polygons — 112
- 7.1 Polygonal map quadtrees — 112
  - 7.1.1 PM1 quadtrees — 116
  - 7.1.2 PM2 quadtrees — 122
  - 7.1.3 PM3 quadtrees — 125
- 7.2 R-trees — 126
- 7.3 Notes — 136
- 7.4 Exercises — 136

III SPATIAL ANALYSIS AND MODELING — 137

8 Interpolation — 139
- 8.1 Inverse distance weighted interpolation — 141
- 8.2 Kriging — 146
  - 8.2.1 Semivariance — 147
  - 8.2.2 Modeling semivariance — 150
  - 8.2.3 Ordinary kriging — 156
  - 8.2.4 Simple kriging — 162
- 8.3 Using interpolation methods — 165
- 8.4 Midpoint displacement — 170
- 8.5 Notes — 174
- 8.6 Exercises — 176

9 Spatial Pattern and Analysis — 177
- 9.1 Point pattern analysis — 178
  - 9.1.1 Nearest neighbor analysis — 178
  - 9.1.2 Ripley's *K*-function — 185
- 9.2 Spatial autocorrelation — 191
- 9.3 Clustering — 199
- 9.4 Landscape ecology metrics — 202
- 9.5 Notes — 208
- 9.6 Exercises — 209

Contents

10 Network Analysis — **211**
- 10.1 Network traversals — 214
  - 10.1.1 Breadth-first traversal — 214
  - 10.1.2 Depth-first traversal — 216
- 10.2 Single source shortest path — 217
- 10.3 All pair shortest paths — 222
- 10.4 Notes — 225
- 10.5 Exercises — 226

11 Spatial Optimization — **228**
- 11.1 1-center location problem — 230
- 11.2 Location problems — 244
- 11.3 Notes — 248
- 11.4 Exercises — 249

12 Heuristic Search Algorithms — **251**
- 12.1 Greedy algorithms — 251
- 12.2 Vertex exchange algorithm — 253
- 12.3 Simulated annealing — 261
- 12.4 Notes — 273
- 12.5 Exercises — 274

Postscript — **275**

Appendix

A Python: A Primer — **277**
- A.1 List comprehension — 280
- A.2 Functions, modules, and recursive functions — 282
- A.3 Lambda functions and sorting — 283
- A.4 NumPy and Matplotlib — 284
- A.5 Classes — 288

B GDAL/OGR and PySAL — **291**
- B.1 OGR — 292
  - B.1.1 Attributes — 293
  - B.1.2 Geometry and coordinates — 293
  - B.1.3 Projecting points — 294
  - B.1.4 Projecting and writing geospatial data — 295
  - B.1.5 Adjacency matrix — 298
- B.2 GDAL — 300
- B.3 PySAL — 301

C Code List — **303**

References — **307**
Index — **314**

About the Author

Ningchuan Xiao is an associate professor in the Department of Geography at the Ohio State University. He has taught a wide range of courses in cartography, GIS, and spatial analysis and modeling. He previously served as chair of the Spatial Analysis and Modeling Specialty Group of the Association of American Geographers from 2009 to 2012. Dr Xiao's research focuses on the development of effective and efficient computational methods for mapping and analyzing spatial and temporal data in various application domains, including spatial optimization, spatial decision support systems, environmental and ecological modeling, and spatial epidemiology. His research has been published in leading journals in geography and GIScience. His current projects include designing and implementing novel approaches to analyzing and mapping big data from social media and other online sources, and developing search algorithms for solving spatial aggregation problems. He is also working with interdisciplinary teams on projects to map the impacts of human mobility on transmission of infectious diseases and to model the impacts of environment on social dynamics in the Far North Region of Cameroon.

Preface

Geographic information systems (GIS) have become increasingly important in helping us understand complex social, economic, and natural dynamics where spatial components play a key role. In the relatively short history of GIS development and applications, one can often observe an alienating process that separates GIS as a "machine" from its human users. In some cases, GIS have been reduced to a black box that can be used to generate pleasant mapping products serving various purposes. This trend has already encroached on our GIS education programs where a noticeable portion of teaching has focused on training students how to use the graphical user interface of GIS packages. Addressing this situation presents a great challenge to GIS researchers and educators.

In this book we focus on critical algorithms used in GIS that serve as a cornerstone in supporting the many operations on spatial data. The field of GIS has always been diverse, and many textbooks typically cover the general topics without providing in-depth discussion for students to fully understand the context of GIS concepts. Algorithms in GIS are often presented in different ways using different data structures, and the lack of a coherent representation has made it difficult for students to digest the essence of algorithms. Students in GIS classes often come from a variety of research and educational backgrounds and may not be fully comfortable with the terms used in traditional and formal algorithm description. All of these have seemingly made algorithms a difficult topic to teach in GIS classes. But they should not be excuses for us to avoid algorithm topics. By examining how spatial data are input into an algorithm and how the algorithm is used to process the data to yield the output, we can gain substantial understanding of two important components of GIS: what geospatial data actually are and how these data are actually processed.

This book covers algorithms that are critical in implementing some major GIS functions. The goal, however, is not to go over an exhaustive and long list of algorithms. Instead, we only include algorithms that either are commonly used in today's GIS or have a significant influence on the development of the current algorithms; as such the choice of topics may appear to be subjective. We look at geospatial data from a minimalist perspective by simply viewing them as being spatial and locational, meaning that we focus on the coordinates using an atomic view of geographic information where most geospatial data can be understood as collections of points. In this sense, we boil down a lot of unnecessary discussion about the difference between vector and raster to a fundamental data model. Starting from there, we dive into the diversity of GIS algorithms that can be used to help us carry out some fundamental functions: measuring important spatial properties such as distance, incorporating multiple data sources using overlay, and speeding up analysis using various indexing techniques. We also go to certain lengths to cover algorithms for spatial analysis and modeling tasks such as interpolation, pattern analysis, and decision-making using optimization models. Admittedly all these functions are available in many GIS packages, open source or commercial. However, our goal is not to duplicate what is available out there, but to show how things out there work so that we can implement our own GIS, or at least GIS functions, without relying on a software system labeled as "GIS." To gain such freedom, we must go to the level where data are transparent not packaged, processes are outlined as code not as buttons to click on, and results are as transparent as input.

This is not a traditional book on algorithms. In a typical computer science book one would expect algorithms to be presented using pseudo-code that is assumed to concentrate on the critical parts of algorithms but to ignore some of the details. The pseudo-code approach, however, is not suitable for many of those who study GIS because the theoretical aspects of the algorithms are often not the main concern. Instead, the drive to understand algorithms in GIS typically comes from the desire to know "how stuff works." For this reason, real, working code is used in this book to present and implement the algorithms. We specifically use Python as the language in this book mainly because of its simplistic programming style that eliminates much of the overhead in other programming languages. Careful thought has been given to balancing the use of powerful but otherwise "packaged" Python modules in order to show the actual mechanism in the algorithms. With that, we also maintain a coherent geospatial data representation and therefore data structures, starting from the beginning of the book.

This book would not be in the form it is without help. I am deeply grateful to the open source community that has tremendously advanced the way we think about and use spatial data today, and in this sense, this book would not even exist without open source developments. I have on countless occasions consulted with the Stack Overflow sites for my Python questions. The online LaTeX community (http://tex.stackexchange.com and http://en.wikibooks.org/wiki/LaTeX) always had answers to my typesetting questions during the many all-nighters spent writing this book. Some open source packages are especially crucial to many parts of the book, including those on *k*-D trees (http://en.wikipedia.org/wiki/K-d_tree and https://code.google.com/p/python-kdtree/) and kriging (https://github.com/cjohnson318/geostatsmodels). My thanks are due to students in many of the classes I have taught in the past few years in the United States and in China. Their feedback and sometimes critiques have enabled me to improve my implementations of many of the algorithms. I thank Mei-Po Kwan for her support, and Richard Leigh, Katherine Haw and Matthew Oldfield for carefully handling the manuscript. Detailed comments from a number of reviewers of an earlier version of this book have greatly helped me enhance the text. Finally, I would like to thank my family for their patience and support during the writing process.

Ningchuan Xiao
−82.8650°, 40.0556°
December 2014

1

Introduction

Algorithms[1] are designed to solve computational problems. In general, an algorithm is a process that contains a set of well designed steps for calculation. For example, to correctly calculate the sum of 18 and 19, one must know how to deal with the fact that 8 plus 9 is more than 10, though different cultures have different ways of processing that. Even for a simple problem like this, we expect that the steps used can help us get the answer quickly and correctly. There are many problems that are more difficult than the simple problem of addition, and solving these problems, again efficiently and correctly, requires more careful design of the computational steps.

In GIS development and applications, algorithms are important in almost every aspect. When we click on a map, for example, we expect a quick response from the computer system so that we can pull out relevant information about the point or area we just clicked on. Such a fundamental daily routine for almost every GIS application involves a variety of algorithms to ensure a satisfying response. It starts from searching for the object (point, line, polygon, or pixel) underneath the clicking point. An efficient search algorithm will allow us to narrow down to the area of interest quickly. While a brute-force approach may work by physically checking every object in our data, it will not be useful for a large data set that will make the time of finding the target impractically long. Many spatial indexing and query algorithms are designed to address this issue. While the search is ongoing, we must check whether the object in our data matches the point clicked. For polygons, we must decide whether the click point is within a polygon in our data, which requires a special algorithm to quickly return a yes or no answer to decide whether the point is in the polygon. Geospatial data normally come from many different sources, and it has been common practice to transform them into the same coordinate system so that different data sets can be processed consistently. Another common application of the multiple data sources is to overlay them to make the information more useful together.

There are many aspects of an algorithm to be examined. It is straightforward to require that an algorithm solve the problem correctly. For some algorithms, it is easy to prove their

---

[1]The word "algorithm" itself comes from medieval Latin *algorismus*, which is from the name of *al-Khwārizmī*, a Persian mathematician, astronomer, and geographer, who made great contributions to the knowledge of algebra and world geography.

correctness. For example, we will introduce two search algorithms later in this chapter, and their correctness should be quite straightforward. Other algorithms, however, are not so obvious, and proving their correctness will require more formal analysis. A second feature of algorithms is their efficiency or running time. Of course we always want an algorithm to be fast, but there are theoretical limits on how fast or efficient an algorithm can be, as determined by the problem. We will discuss some of those problems at the end of the book under topics of spatial optimization. Besides correctness and running time, algorithms are often closely related to how the data are organized to enable the processes and how the algorithms are actually implemented.

1.1 Computational concerns for algorithms

Let us assume we have a list of n points and the list does not have any order. We want to find a point from the list. How long will we take to find the point? This is a reasonable question. But the actual *time* is highly related to a lot of issues such as the programming language, the skill of the person who codes the program, the platform, the speed and number of the CPUs, and so on. A more useful way to examine the time issue is to know how many *steps* we need to finish the job, and then we analyze the total cost of performing the algorithm in terms of the number of steps used. The cost of each step is of course variable and is dependent on what constitutes a step. Nevertheless, it is still a more reliable way of thinking about computing time because many computational steps, such as simple arithmetic operations, logical expressions, accessing computer memory for information retrieval, and variable value assignment, can be identified and they only cost a constant amount of time. If we can figure out a way to count how many steps are needed to carry out a procedure, we will then have a pretty good idea about how much time the entire procedure will cost, especially when we compare algorithms.

Returning to our list of points, if there is no structure in the list – the points are stored in an arbitrary order – the best we can do to find a point from the list is to test all the points in the list, one by one, until we can conclude it is in or not in the list. Let us assume the name of the list is `points` and we want to find if the list includes point `p0`. We can use a simple algorithm to do the search (Listing 1.1).

Listing 1.1: Linear search to find point `p0` in a list.

```
1  for each point p in points:
2      if p is the same as p0:
3          return p and stop
```

The algorithm in Listing 1.1 is called a linear search; in it we simply go through all the points, if necessary, to search for the information we need. How many steps are necessary in this algorithm? The first line is a loop and, because of the size of the list, it will run as many as n times when the item we are looking for happens to be the last one in the list. The cost of running just once in the loop part in line 1 is a constant because the list is stored in the computer memory, and the main operation steps here are to access the information at a fixed location in the memory and then to move on to the next item in the memory. Suppose that the cost is c_1 and we will run it up to to n times in the loop. The second line is a logic comparison between two points. It will run up to n times as well because it is inside the loop. Suppose that the cost of doing a logic comparison is c_2 and it is a constant too. Line 3 simply returns the value of the point found; it has a constant cost of c and it will only run once. For the best case scenario, we will find the target at the first iteration of the loop and therefore the total cost is simply $c_1 + c_2 + c$, which can be generalized as a constant $b + c$. In the worst case scenario, however, we will need to run all the way to the last item in the list and therefore the total cost becomes $c_1 n + c_2 n + c$, which can be generalized as $bn + c$, where b and c are constants, and n is the size of the list (also the size of the problem). On average, if the list is a random set of points and we are going to search for a random point many times, we should expect a cost of $c_1 n/2 + c_2 n/2 + c$, which can be generalized as $b'n + c$, and we know $b' < b$, meaning that it will not cost as much as the worse case scenario does.

How much are we interested in the actual values of b, b', and c in the above analysis? How will these values impact the total computation cost? As it turns out, not much, because they are constants. But adding them up many times will have a real impact and n, the problem size, generally controls how many times these constant costs will be added together. When n reaches a certain level, the impact of the constants will become minimal and it is really the magnitude of n that controls the *growth* of the total computation cost.

Some algorithms will have a cost related to n^2, which is significantly different from the cost of n. For example, the algorithm in Listing 1.2 is a simple procedure to compute the shortest pairwise distance between two points in the list of n points. Here, the first loop (line 2) will run n times at a cost of t_1 each, and the second loop (line 3) will run exactly n^2 times at the same cost of t_1 each. The logic comparison (line 4) will run n^2 times and we assume each time the cost is t_2. The calculation of distance (line 5) will definitely be more costly than the other simple operations such as logic comparison, but it is still a constant as the input is fixed (with two points) and only a small finite set of steps will be taken to carry out the calculation. We say the cost of each distance calculation is a constant t_3. Since we do not compute the distance between the point and itself, the distance calculation will run $n^2 - n$ times, as will the comparison in line 6 (with time t_4). The assignment in line 7 will cost a constant time of t_5 and may run up to $n^2 - n$ times in the worst case scenario where every next distance is shorter than the previous one. The last line will only run once with a time of c. Overall, the total time for this algorithm will be $t_1 n + t_1 n^2 + t_2 n^2 + t_3 (n^2 - n) + t_4 (n^2 - n) + t_5 (n^2 - n) + c$, which can be generalized as $an^2 + bn + c$. Now it should be clear that this algorithm has a running time that is controlled by n^2.

Listing 1.2: Linear search to find shortest pairwise distance in a list of points.

```
1  let mindist be a very large number
2  for each point p1 in points:
3      for each point p2 in points:
4          if p1 is not p2:
5              let d be the distance between p1 and p2
6              if d < mindist:
7                  mindist = d
8  return mindist and stop
```

In the two example algorithms we have examined so far, the order of n indicates the total cost and we say that our linear search algorithm has a computation cost in the order of n and the shortest pairwise distance algorithm in the order of n^2. For the linear search, we also know that, when n increases, the total cost of search will always have an upper bound of bn. But is there a lower bound? We know the best case scenario has a running time of a constant, or in the order of n^0, but that does not apply to the general case. When we can definitely find an upper bound but not a lower bound of the running time, we use the O-notation to denote the order. In our case, we have $O(n)$ for the average case, and the worst case scenario as well (because again the constants do not control the total cost). In other words, we say that the running time, or time complexity, of the linear search algorithm is $O(n)$. Because the O-notation is about the upper bound, which is meant to be the worst case scenario, we also mean the time complexity of the worst case scenario.

There are algorithms for which we do not have the upper bound of their running time. But we know their lower bound and we use the Ω-notation to indicate that. A running time of $\Omega(n)$ would mean we know the algorithm will at least cost an order of n in its running time, though we do not know the upper bound of the running time. For other algorithms, we know both upper and lower bounds of the running time and we use the Θ-notation to indicate that. For example, a running time of $\Theta(n^2)$ indicates that the algorithm will take an order of n^2 in running time in all cases, best and worst. This is the case for our shortest distance algorithm because the process will always run n^2 times, regardless of the outcome of the comparison in line 6. It is more accurate to say the time complexity is $\Theta(n^2)$ instead of $O(n^2)$ because we know the lower bound of the running time of pairwise shortest distance is always in the order of n^2.

Now we reorganize our points in the previous list in a particular tree structure as illustrated in Figure 1.1. This is a binary tree because each node on the tree can have at most two branches, starting from the root. Here the root of the tree stores point (6, 7) and we show it at the top. All the points with X coordinates smaller than or equal to that at the root are stored in the left branches of the root and those with X coordinates greater than that of the root point are stored in the right branches of the root. Going down the tree to the second level, we have two points there, (4, 6) and (9, 4). For each of these points, we make sure that the rest of the points will be stored on the left if they have a smaller or equal Y coordinate value, and on the right if greater. We alternate the use of X and Y coordinates going down the tree, until we find the point we are looking for or reach the end of a branch (a leaf node).

To use the tree structure to search for a point, we start from the root (we always start from the root for a tree structure) and go down the tree by determining which branch to proceed along using the appropriate coordinate at each level of the tree. For example, to

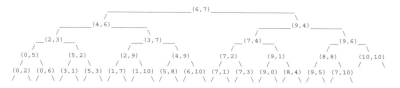

Figure 1.1 A tree structure that stores 29 random points. Each node of the tree is labeled using a point with X and Y coordinates that range from 0 to 10

search for a target point of (1, 7), we first go to the left branch of the root because the X coordinate of our target point is 1, smaller than that in the root. Then we go the the right branch of the second level node of (4, 6) because the Y coordinate (7) is greater than that in the node. Now we reach the node of (3, 7) at the third level of the tree and we will go to the left branch there because the target X coordinate is smaller than that in the node. Finally, we reach the node (2, 9) and we will move to its left branch because the Y coordinate in the target is smaller than that in the node. In sum, given a tree, we can write the algorithm in Listing 1.3 to fulfill such a search strategy using a tree structure.

Listing 1.3: Binary search to find point p0 in a tree.

```
let t be the root of the tree
while t is not empty:
    let p be the point at node t
    if p is the same as point p0:
        return p and stop
    if t is on an even level of the tree:
        coordp, coordp0 = X coordinates of p and p0
    else:
        coordp, coordp0 = Y coordinates of p and p0
    if coordp0 <= coordp:
        t = the left branch of t
    else:
        t = the right branch of t
```

This is called a binary search, using the tree structure. Based on our discussion about running time, it is straightforward to see that the running time of this search algorithm is determined by the number of times we have to run the while loop (line 2), which is determined by the height of the tree, as defined by the number of edges from the root to the farthest leaf node. The above tree has a height of 4 and can hold up to 31 points (we only have 29 here). In general, for a binary tree with a height of H, we can store up to $2^0 + 2^1 + 2^2 + \ldots + 2^H = 2^{H+1} - 1$ items. In other words, if we have n points in total that fill all the nodes in a perfectly balanced binary tree where all the leaf nodes are at exactly the same level, we have $2^{H+1} - 1 = n$ and hence $H = \log_2(n + 1) - 1$. In this case, when

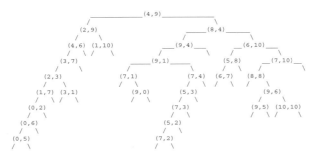

Figure 1.2 An unbalanced tree structure to store 29 random points

such a tree is given, we have a running time of the order of $\log_2(n+1)$, which is at the same order as $\log_2 n$, and we say the running time is $O(\log_2 n)$. For a balanced but not perfect tree, meaning the difference between the heights of all leaf nodes is at most 1, we can still achieve a running time of $O(\log_2 n)$ since the farthest leaf node has a height of H. When a balanced tree cannot be guaranteed, however, things can get worse. An example of an unbalanced tree is given in Figure 1.2 where we have exactly the same points but the tree has a height of 8. Therefore, we know the binary search algorithm is more efficient than linear search because $O(\log_2 n) < O(n)$, but the actual running time is dependent on how the tree is constructed and can be longer than $O(\log_2 n)$ if the tree is unbalanced.

Since using a balanced binary tree to store the points can greatly improve the efficiency of search, will that also help to reduce the running time of calculating the shortest pairwise distance? The brute-force approach we discussed here has a running time of $\Theta(n^2)$, which grows quickly as n increases. It would be reasonable to look for a more efficient way to get the shortest pairwise distance in a list of points. The answer to our question is yes, and the key lies in the use of tree structures. We will continue to explore this topic in the second part of the book where we discuss different tree structures in more depth. Throughout the book, we do not focus much on the theoretical analysis of running time for each algorithm. Instead, we will conduct more empirical analysis by actually running the algorithms on different data sets.

The discussion about search and especially binary search on a tree logically leads to the topic of data structure: how we store and organize data to facilitate the procedures in an algorithm. A tree structure is a good example of how the original data stored in a list can be reorganized to achieve better search performance. Many data structures are problem-specific. Some data structures can be complicated, but the increase in storage is often compensated by a decrease in running time.

1.2 Coding

Algorithms can be described in different ways. We use verbal statements in this chapter to describe the linear and binary search algorithms. For theoretical work, a formal description that details the steps but is not necessarily executable will suffice. We call this type of description pseudo-code because it is not real computer code, though very close. In this book, we take a more practical and explicit route by describing algorithms in an actual computer programming language. Specifically, we use Python to describe the algorithms covered in this book.

Writing computer programs (i.e., coding) to describe algorithms has a substantial benefit: all the algorithms will immediately be executable. In this way, we present everything related to how the algorithms work in the plain text of the book. The code becomes part of the text and consequently becomes an open source experiment where each line of the process can be examined, modified, and improved. However, this code as text approach may present too much information, especially when the programming language may need ancillary code to help the main task. For example, many programming languages require end of line symbols and brackets as part of the code to ensure syntax correctness. These symbols, when added as part of the text, may hamper the reading process and therefore make it difficult for us to concentrate on the main contents of the text. We choose Python in this book largely because of its simple syntax, along with many of its popular, powerful, and well maintained modules. All the programs listed in this book were tested in Python 2.7, which is a stable and widely adopted version at the time of writing. The majority of the programs in this book only use the basic Python features, so these programs in principle are likely to be compatible with newer versions of Python.

Python has become a popular programming language in recent years, including the use of Python for plugins in GIS packages such as QGIS and ArcGIS. It is important to point out that programming in Python is a skill that can definitely be acquired through learning and practice. To help readers make a quick start on the language, a short introduction to Python is included as Appendix A. This is not a comprehensive tutorial as many of the online tutorials have more detailed and in-depth discussion about the language. However, many of Python's useful features, especially those related to the main text, are included in the tutorial.

1.3 How to use this book

The main text of this book is divided into three major parts. The general logic here is to start the book with a discussion on the most fundamental aspects of the data – the geometry – before we move on to more advanced topics in spatial indexing and spatial analysis and modeling. At the end of each chapter we review the major literature related to the topics covered in a section called Notes. At the end of the book, we also include three appendices to help readers understand the Python programming language and the structure of the programs included in the book.

In Part I, we focus on locations, or more specifically on coordinates that can be used to help us understand geospatial information. In Chapter 2, we examine a few algorithms to compute different kinds of distance, such as distances between points and distance from a point to a line. We also look at the calculation of polygon centroids and a widely used algorithm called point-in-polygon that efficiently helps us determine whether a point is located within a polygon. The final topic in Chapter 2 is about the transformation between coordinate systems involving map projections. Chapter 3 covers a traditional GIS operation, known as overlay. As "old" as this topic is, the actual computation of overlaying two polygons can be tedious, though not necessarily complicated. Many of the topics in this part of the book are related to the field of computational geometry. But we focus on those that are most relevant to the GIS world.

Part II is centered around the idea of spatial indexing. Spatial information is special. Though the general concept in indexing, divide and conquer, is the same for spatial information, because of the two dimensions (or more in some cases) in spatial information, more dedicated algorithms must be designed. We first introduce the basic concepts of indexing in Chapter 4, where we focus on the development of a tree structure. Chapter 5 is devoted to k-D trees that are commonly used to index point data. Chapter 6 covers a popular indexing technique called quadtrees, for both point and raster data. Chapter 7 extends the discussion to indexing lines and polygons in spatial data.

Part III of the book focuses on the heart of GIS applications: spatial analysis and modeling. We first explore the interpolation methods on point data in Chapter 8 where we compare and contrast two commonly used interpolation methods: inverse distance weighting and kriging. We also include a data simulation algorithm called midpoint displacement from the fractal geometry literature. Chapter 9 is devoted to spatial pattern analysis where the calculation of indices such as Moran's I is included. Algorithms for network analysis, especially those calculating the shortest paths, are included in Chapter 10. We devote two chapters to topics in spatial optimization: in Chapter 11 we focus on the exact methods and in Chapter 12 we explore some of the heuristic methods.

In addition to the three parts in the main text, we have also included three appendices to cover some of the technical details about coding. It should be obvious that, while we talk about algorithms most of the time, this book is about coding as well. For this reason, a short introduction to Python is first included. Then we compile a short introduction on the Python binding of a powerful library called GDAL/OGR and a Python library for spatial analysis called PySAL. The purpose here is to help readers quickly get started with these libraries so that they can have a sense how "real-world" data sets can be closely related to the topics (and code of course) presented in this book.

Most of the programs listed in this book are also given a file name that can be used in other programs. In that case the name of the program is listed in the caption of each code listing. We use directories to organize the programs. In the last appendix, we provide an overview of the code by listing all the Python programs and example data sets and discuss how they can be used.

Each of the chapters in this book could easily be extended into another book to cover the depth of each topic. This book, then, is a survey of the general topics in GIS algorithms. The best way to grasp the breadth of the topics presented in this book is coding. A collective Github page[2] is under development where readers can contribute their thoughts on implementing the algorithms included in this book and new algorithms that are beyond the scope of the book. While theoretical derivation is not the focus of this book, empirical analysis definitely is. We have included many experiments in the text and have suggested more in the exercises. This, however, should not limit further experiments, especially those that are innovative and overarch various topics. The book will only be successful if it achieves two goals: first, based on the skills built on understanding the algorithms and code, that readers are able to develop their own tool sets that fit different data sets and application requirements; and second, that coding becomes a habit when it comes to dealing with geospatial data.

---

[2] https://github.com/gisalgs

Part I

Geometric Algorithms

2

Basic Geometric Operations

Imagine a vast sheet of paper on which straight Lines, Triangles, Squares, Pentagons, Hexagons, and other figures, instead of remaining fixed in their places, move freely about, on or in the surface, but without the power of rising above or sinking below it, very much like shadows – only hard with luminous edges – and you will then have a pretty correct notion of my country and countrymen.

Edwin A. Abbott, *Flatland: A Romance of Many Dimensions*

This chapter focuses on the essential algorithms in GIS that involve geometric operations. We first introduce the calculation of distance between two points and then move on to geometric objects with higher dimensions, including algorithms to determine properties such as point-in-polygon, point distance to lines, polygon centroid, and polygon area. We will also examine two map projections and discuss how to transform geospatial data from one coordinate system to another.

2.1 Point

Before we start our discussion, let us define a data structure for a point that will be used throughout this book. This is a Python class called `Point` (Listing 2.1). We are only interested in two-dimensional cases and we just store the X and Y coordinates of a point in the class. We override a few Python built-in methods to provide some convenient features. The `__getitem__` method allows us to iterate through the X and Y coordinates of a point using indices 0 and 1, respectively. The `__len__` method returns the number of dimensions in the point (we only return two here). We should also be able to judge whether two points are identical if they have exactly the same X and Y coordinates (`__eq__`), or different if the coordinates are different (`__ne__`). Additionally, we also need a way to compare points in a coordinate system so that a point on the lower left-hand side of the coordinate system is always "smaller" than those on the upper right-hand side. For two points p_1 and p_2, we say that $p_1 < p_2$ if p_1 has a smaller X coordinate. For two points with the same X coordinate, the lower point with a smaller Y coordinate is considered as smaller. This is captured by overriding the built-in comparison operators of `__lt__` (less than), `__gt__` (greater than), `__le__` (less than or equal to), and `__ge__` (greater than or equal to). We also use the coordinates of the point when it is printed out in text (`__str__` and `__repr__`). While we do not specifically use this type of ordering and comparison in this chapter, they play a big role in later chapters. A `distance` method is included to compute the Euclidean distance between two points.

Listing 2.1: Data structure for a point class (point.py).

```python
from math import sqrt
class Point():
    """A class for points in Cartesian coordinate systems."""
    def __init__(self, x=None, y=None):
        self.x, self.y = x, y
    def __getitem__(self, i):
        if i==0: return self.x
        if i==1: return self.y
        return None
    def __len__(self):
        return 2
    def __eq__(self, other):
        if isinstance(other, Point):
            return self.x==other.x and self.y==other.y
        return NotImplemented
    def __ne__(self, other):
        result = self.__eq__(other)
        if result is NotImplemented:
            return result
        return not result
    def __lt__(self, other):
        if isinstance(other, Point):
            if self.x<other.x:
                return True
            elif self.x==other.x and self.y<other.y:
                return True
            return False
        return NotImplemented
    def __gt__(self, other):
        if isinstance(other, Point):
            if self.x>other.x:
                return True
            elif self.x==other.x and self.y>other.y:
                return True
            return False
        return NotImplemented
    def __ge__(self, other):
        if isinstance(other, Point):
            if self > other or self == other:
                return True
            else:
                return False
        return False
```

```
44            return NotImplemented
45        def __le__(self, other):
46            if isinstance(other, Point):
47                if self < other or self == other:
48                    return True
49                else:
50                    return False
51                return False
52            return NotImplemented
53        def __str__(self):
54            if type(self.x) is int and type(self.y) is int:
55                return "({0},{1})".format(self.x,self.y)
56            else:
57                return "({0:.1f}, {1:.1f})".format(self.x,self.y)
58        def __repr__(self):
59            if type(self.x) is int and type(self.y) is int:
60                return "({0},{1})".format(self.x,self.y)
61            else:
62                return "({0:.1f}, {1:.1f})".format(self.x,self.y)
63        def distance(self, other):
64            return sqrt((self.x-other.x)**2 + (self.y-other.y)**2)
```

The Point class can be easily extended to become a more flexible representation of points in geospatial data. For example, while we assume X and Y are on a Cartesian plane (because of the way the distance method is defined), we can relax this constraint by adding a new data member of the class called CS that can be used to specify the kind of coordinate system for the point and we can calculate the distance accordingly. We can create a subclass that inherits from the Point class so that points in higher dimensions can be represented. We can also add time and other attributes to the class.

2.2 Distance between two points

The distance between two points can be calculated in various ways, largely depending on the application domain and the coordinate system in which the two points are measured. The most commonly used distance measure is the Euclidean distance on the straight line between the two points on a Cartesian plane where the distance can be simply computed as

$$d = \sqrt{(x_1 - x_2)^2 + (y_1 - y_2)^2},$$

where x_1 and x_2 are the horizontal (X) coordinates of the two points and y_1 and y_2 are their vertical (Y) coordinates. The distance method in the Point class returns the Euclidean distance between two points (Listing 2.1).

In some special cases, the Euclidean distance may not be a suitable measure between two points, especially in an urban setting where one must follow the street network to go from one point to another (Figure 2.1). In this case, we can use the Manhattan distance between these two points:

$$d_1 = |x_1 - x_2| + |y_1 - y_2|.$$

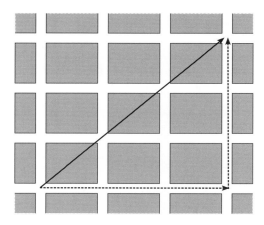

Figure 2.1 Euclidean (solid line) and Manhattan distance (dashed lines). Shaded areas represent buildings in an urban setting where roads run between buildings

When the coordinates of the points are measured on a spherical surface, the above distance measures will not give the correct distance between the points. In this case, the shortest distance between the two points is measured on the great circle that passes through the two points. More specifically, when each point is measured by its latitude or the angle to the equatorial plane (φ) and longitude or the angle between the meridian of the point and the prime meridian (λ), the arc distance α (angle) between two points on the great circle is given by

$$\cos\alpha = \sin\varphi_1 \sin\varphi_2 + \cos\varphi_1 \cos\varphi_2 \cos d\lambda,$$

where $d\lambda = |\lambda_1 - \lambda_2|$ is the absolute difference between the longitudes of the two points. In the actual calculation, however, we use the following Haversine formula to avoid the difficulty of handling negative values:

$$a = \sin^2\frac{d\varphi}{2} + \cos\varphi_1 \cos\varphi_2 \sin^2\frac{d\lambda}{2},$$

$$c = 2\arcsin\min(1, \sqrt{a}),$$

where the min function returns the smaller value of the two inputs in order to avoid numerical artifacts that may cause the value of a to be greater than 1; then the distance can be calculated as

$$d = cR,$$

where R is the radius of the spherical earth (3,959 miles or 6,371 kilometers). It is important to note that the latitude and longitude readings must be converted to radians before the above formula can be used to calculate the distance.

We now write a Python program to calculate the spherical distance between two points (Listing 2.2). Here the latitudes and longitudes are provided in degrees and we convert them to radians (lines 13–16). Using this code, we can calculate the distance between Columbus, OH (40° N, 83° W) and Beijing (39.91° N, 116.56° E) as 6,780 miles (10,911 kilometers).

Listing 2.2: Calculating the great circle distance (spherical_distance.py).

```
import math
def spdist(lat1, lon1, lat2, lon2):
    """
    Calculates the great circle distance given
    the latitudes and longitudes of two points.
    Input
      lat1, lon1: lat and long in degrees for the first point
      lat2, lon2: lat and long in degrees for the second point
    Output
      d: great circle distance
    """
    D = 3959                           # earth radius in miles
    phi1 = math.radians(lat1)
    lambda1 = math.radians(lon1)
    phi2 = math.radians(lat2)
    lambda2 = math.radians(lon2)
    dlambda = lambda2 - lambda1
    dphi = phi2 - phi1
    sinlat = math.sin(dphi/2.0)
    sinlong = math.sin(dlambda/2.0)
    alpha=(sinlat*sinlat) + math.cos(phi1) * \
        math.cos(phi2) * (sinlong*sinlong)
    c=2 * math.asin(min(1, math.sqrt(alpha)))
    d=D*c
    return d

if __name__ == "__main__":
    lat1, lon1 = 40, -83               # Columbus, OH
    lat2, lon2 = 39.91, 116.56         # Beijing
    print spdist(lat1, lon1, lat2, lon2)
```

2.3 Distance from a point to a line

Let $ax+by+c=0$ be a line on the plane, where a, b, and c are constants. The distance from a point (x_0, y_0) on the plane to the line is computed as

$$\frac{|ax_0 + by_0 + c|}{\sqrt{a^2 + b^2}}.$$

We can prove this by calculating the distance from the point to the intersection point between line $ax+by+c=0$ and its perpendicular line that goes through point (x_0, y_0). Let (x_1, y_1) be the intersection point, and we know the slope of the perpendicular line is b/a. Hence, we have

$$\frac{y_1 - y_0}{x_1 - x_0} = \frac{b}{a},$$

which gives

$$a(y_1 - y_0) - b(x_1 - x_0) = 0.$$

We then take the square of both sides and rearrange the result to give

$$a^2(y_1 - y_0)^2 + b^2(x_1 - x_0)^2 = 2ab(x_1 - x_0)(y_1 - y_0).$$

We add $a^2(x_1 - x_0)^2 + b^2(y_1 - y_0)^2$ to both sides. The left-hand side can be rewritten as

$$(a^2 + b^2)[(y_1 - y_0)^2 + (x_1 - x_0)^2],$$

and the right-hand side as

$$[a(x_1 - x_0) + b(y_1 - y_0)]^2 = [ax_1 + by_1 - ax_0 - by_0]^2.$$

Because point (x_1, y_1) is also on the original line, we have $ax_1 + by_1 = -c$. Hence

$$(a^2 + b^2)[(y_1 - y_0)^2 + (x_1 - x_0)^2] = [ax_0 + by_0 + c]^2.$$

The term $(y_1 - y_0)^2 + (x_1 - x_0)^2$ is the square of the distance between (x_0, y_0) and line $ax + by + c$. We then have the distance as

$$\sqrt{(y_1 - y_0)^2 + (x_1 - x_0)^2} = \frac{|ax_0 + by_0 + c|}{\sqrt{a^2 + b^2}}.$$

In most GIS applications, we can easily get information about a line segment defined by the two endpoints. It is therefore necessary to compute the values of a, b, and c in the above equations using the two points. Let (x_1, y_1) and (x_2, y_2) be the two endpoints. We define $dx = x_1 - x_2$ and $dy = y_1 - y_2$. We can then write the line equation using the slope:

$$y = \frac{dy}{dx}x + n,$$

where n is a constant that we will compute. We now plug in the first endpoint to get

$$y_1 = \frac{dy}{dx}x_1 + n,$$

which gives

$$n = y_1 - \frac{dy}{dx}x_1.$$

Therefore, we have a general form for the line as

$$y = \frac{dy}{dx}x + y_1 - \frac{dy}{dx}x_1.$$

We multiply both sides of the above equation by dx and rearrange it as

$$xdy - ydx + y_1 dx - x_1 dy = 0.$$

Now we have $a = dy$, $b = -dx$, and $c = y_1 dx - x_1 dy$. With these parameters, we can quickly calculate the distance between a point and the line segment (Listing 2.3).

Listing 2.3: A Python program to calculate point to line distance (point2line.py).

```
import math
from point import *

def point2line(p, p1, p2):
    """
    Calculate the distance from point to a line.
    Input
      p: the point
      p1 and p2: the two points that define a line
    Output
      d: distance from p to line p1p2
    """
    x0 = float(p.x)
    y0 = float(p.y)
    x1 = float(p1.x)
    y1 = float(p1.y)
    x2 = float(p2.x)
    y2 = float(p2.y)
    dx = x1-x2
    dy = y1-y2
    a = dy
    b = -dx
    c = y1*dx - x1*dy
    if a==0 and b==0:            # p1 and p2 are the same point
        d = math.sqrt((x1-x0)*(x1-x0) + (y1-y0)*(y1-y0))
```

```
26        else:
27            d = abs(a*x0+b*y0+c)/math.sqrt(a*a+b*b)
28        return d
29
30 if __name__ == "__main__":
31     p, p1, p2 = Point(10,0), Point(0,100), Point(0,1)
32     print point2line(p, p1, p2)
33     p, p1, p2 = Point(0,10), Point(1000,0.001), Point(-100,0)
34     print point2line(p, p1, p2)
35     p, p1, p2 = Point(0,0), Point(0,10), Point(10,0)
36     print point2line(p, p1, p2)
37     p, p1, p2 = Point(0,0), Point(10,10), Point(10,10)
38     print point2line(p, p1, p2)
```

We use a few test cases (lines 31–38) to demonstrate the use of the code. The output is as follows:

```
10.0
9.9999090909
7.07106781187
14.1421356237
```

2.4 Polygon centroid and area

We represent a polygon P of n points as $(x_1, y_1), (x_2, y_2), \ldots, (x_n, y_n), (x_{n+1}, y_{n+1})$, where we add an additional point in the sequence $(x_{n+1}, y_{n+1}) = (x_1, y_1)$ to ensure that the polygon is closed. If the polygon does not have holes and its boundaries do not intersect, the centroid of the polygon is determined by the coordinates

$$x = \frac{1}{6A} \sum_{i=1}^{n} (x_i + x_{i+1})(x_i y_{i+1} - x_{i+1} y_i),$$

$$y = \frac{1}{6A} \sum_{i=1}^{n} (y_i + y_{i+1})(x_i y_{i+1} - x_{i+1} y_i),$$

where A is the area of the polygon. If the polygon is convex, the centroid is bound to be inside the polygon. However, for concave polygons, the centroid computed as above may be outside the polygon. Though the concept of centroid is related to the center of gravity of the polygon, an outside centroid does not meet the requirement of being the center, which makes it necessary to snap the centroid to a point on the polygon.

To understand how the area formula works, let us use the points of the polygon in Figure 2.2, (a, b, c, d, e, f, g, a), as an example, where a appears twice to ensure polygon closure. Each line segment of the polygon can be used to form a trapezoid,

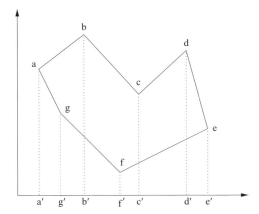

Figure 2.2 Using trapezoids to calculate polygon area. The dotted lines project the nodes of the polygon onto the horizontal axis

and the area of each trapezoid can be calculated as $(x_2 - x_1)(y_2 + y_1)/2$, where subscripts 1 and 2 are used to denote two consecutive points in the sequence. For the case of trapezoid $abb'a'$, the area is calculated as $(x_b - x_a)(y_b + y_a)/2$. For trapezoid $fgg'f'$, the area is $(x_g - x_f)(y_g + y_f)/2$, which will be negative. Clearly, using this formula to calculate the areas of trapezoids underneath line segments ab, bc, cd, and de (note the order of points) will return positive values, while negative values will be obtained for the areas of trapezoids underneath lines ef, fg, and ga. In this way, the formula will finally return the correct area of the polygon by subtracting all the areas enclosed by $efgaa'e'$ from the area enclosed by $abcdee'a'$. The area of the polygon therefore can be computed as

$$A = \frac{1}{2}\sum_{i=1}^{n}(x_{i+1} - x_i)(y_{i+1} + y_i).$$

The above formula is based on the decomposition of a polygon into a series of trapezoids. We can also rewrite the formula into a cross-product form:

$$A = \frac{1}{2}\sum_{i=1}^{n}(x_{i+1}y_i - x_i y_{i+1}).$$

While this cross-product form works exactly the same as the above, in the past it has been more convenient to use for manual calculation. Depending on the order of the points in the sequence, the polygon area formula may return a negative value. Therefore an absolute operation may be needed to obtain the actual area of a polygon.

The Python program in Listing 2.4 returns both the centroid and the area using the above formulas. We also test the program using sample data where the polygon is simply represented using a list of points.

Listing 2.4: A Python program for calculating the area and centroid of a polygon (centroid.py).

```python
from point import *

def centroid(pgon):
    """
    Calculates the centroid and area of a polygon.
    Input
      pgon: a list of Point objects
    Output
      A: the area of the polygon
      C: the centroid of the polygon
    """
    numvert = len(pgon)
    A = 0
    xmean = 0
    ymean = 0
    for i in range(numvert-1):
        ai = pgon[i].x*pgon[i+1].y - pgon[i+1].x*pgon[i].y
        A += ai
        xmean += (pgon[i+1].x+pgon[i].x) * ai
        ymean += (pgon[i+1].y+pgon[i].y) * ai
    A = A/2.0
    C = Point(xmean / (6*A), ymean / (6*A))
    return A, C

# TEST
if __name__ == "__main__":
    points = [ [0,10], [5,0], [10,10], [15,0], [20,10],
               [25,0], [30,20], [40,20], [45,0], [50,50],
               [40,40], [30,50], [25,20], [20,50], [15,10],
               [10,50], [8, 8], [4,50], [0,10] ]
    polygon = [ Point(p[0], p[1]) for p in points ]
    print centroid(polygon)
```

2.5 Determining the position of a point with respect to a line

The calculation of the area of a polygon can return a negative or positive value, depending on the order of the points. This is a very interesting feature and we can use the sign to determine whether a point is on a specific side of a line by considering the area of the triangle of the point and two points that form the line. Let points a, b, and c be the three points on a triangle. Using the cross-product formula, we can obtain the area of this triangle as

$$A(abc) = \frac{1}{2}(x_b y_a - x_a y_b + x_c y_b - x_b y_c + x_a y_c - x_c y_a).$$

Adding $x_b y_b - x_b y_b$ into the parentheses, we can rewrite the above formula in a simpler form:

$$\text{sideplr}(abc) = 2A(abc)$$
$$= (x_a - x_b)(y_c - y_b) - (x_c - x_b)(y_a - y_b).$$

If point a is on the left of the vector formed from point b to c, the above calculation will yield a negative value. For example, in the configurations of points a, b, and c in Figure 2.3A–C, the area of triangle abc (note the order of the points) is calculated in such a way that we always use the area of the trapezoids underneath the triangle (e.g., the trapezoid underneath line cb in 2.3A) to subtract from the total area (e.g., the sum of areas of trapezoids underneath lines ba and ac in Figure 2.3A). This results in a negative value. On the other hand, if a is on the right side of the the vector bc (see Figure 2.3D–F), we will have a positive value. The code for calculating the side is simple, as shown in Listing 2.5.

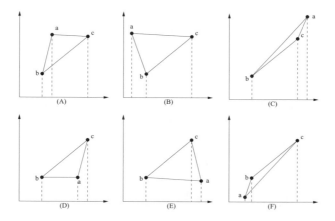

Figure 2.3 The position of point *a* in relation to line *bc*

Listing 2.5: Determining the side of a point (sideplr.py).

```
1   from point import *
2
3   def sideplr(p, p1, p2):
4       """
5       Calculates the side of point p to the vector p1p2.
6       Input
7         p: the point
8         p1, p2: the start and end points of the line
9       Output
10        -1: p is on the left side of p1p2
11        0: p is on the line of p1p2
```

```
            1: p is on the right side of p1p2
        """
        return int((p.x-p1.x)*(p2.y-p1.y)-(p2.x-p1.x)*(p.y-p1.y))

if __name__ == "__main__":
    p=Point(1,1)
    p1=Point(0,0)
    p2=Point(1,0)
    print "Point %s to line %s->%s: %d"%(
        p, p1, p2, sideplr(p, p1, p2))
    print "Point %s to line %s->%s: %d"%(
        p, p2, p1, sideplr(p, p2, p1))
    p = Point(0.5, 0)
    print "Point %s to line %s->%s: %d"%(
        p, p1, p2, sideplr(p, p1, p2))
    print "Point %s to line %s->%s: %d"%(
        p, p2, p1, sideplr(p, p2, p1))
```

Running the test cases in the above code should return the following:

```
Point (1,1) to line (0,0)->(1,0): -1
Point (1,1) to line (1,0)->(0,0): 1
Point (0.5, 0.0) to line (0,0)->(1,0): 0
Point (0.5, 0.0) to line (1,0)->(0,0): 0
```

A collinear alignment of the three points will return a zero. By extending this, we can easily come up with a way to test whether a point is on a line, or lies on one side of the line. We will see how this technique is useful in the next chapter when we discuss the test of intersection between line segments.

2.6 Intersection of two line segments

Let us first consider the intersection between two lines: L_1 passing through points (x_1, y_1) and (x_2, y_2) and L_2 through (x_3, y_3) and (x_4, y_4), with slopes

$$\alpha_1 = \frac{y_2 - y_1}{x_2 - x_1} \quad \text{and} \quad \alpha_2 = \frac{y_4 - y_3}{x_4 - x_3},$$

respectively. Calculating their intersection is straightforward. We first write the equations of the lines for these two segments as

$$y = \alpha_1(x - x_1) + y_1,$$

and

$$y = \alpha_2(x - x_3) + y_3,$$

which can then be used to compute the X coordinate of the intersection point as

$$x = \frac{\alpha_1 x_1 - \alpha_2 x_3 + y_3 - y_1}{\alpha_1 - \alpha_2},$$

and the Y coordinate as

$$y = \alpha_1(x - x_1) + y_1.$$

It is obvious that we will need to consider a few special cases. If the two lines have exactly the same slope, then there will be no intersection. But what if either or both of the lines are vertical with infinite slope? In this case, if both $x_2 - x_1$ and $x_4 - x_3$ are zero then we have two parallel lines. Otherwise, if only one of them is zero, the intersection point has the x coordinate of the vertical line. For example, if $x_1 = x_2$, we have the intersection at point $x = x_1$ and $y = \alpha_2(x_1 - x_3) + y_3$.

When only line segments are considered, however, using the above approach may not be necessary because the two segments may not intersect. We can do a quick check about whether it is possible for the two segments to intersect by looking at their endpoints. If both endpoints of one line segment are on the same side of the other segment, there will be no intersection. We can use the `sideplr` algorithm discussed in the previous section to test the side of a point with respect to a line segment.

Before we formally give the algorithm to compute the intersection between two line segments, let us define a data structure that can be used to effectively store the information about a line segment (Listing 2.6). When we store a segment, we require an edge number (`e`) and the endpoints. We keep a record of the original left point of the line (`lp0` in line 23) because in our later discussion the left point of a line will change due to the calculation of multiple intersection points that move from left to right. Finally, we use the `status` attribute to indicate if the left endpoint of the line segment is the original endpoint (default) or an interior point because of intersection with other segments as a result of polygon overlay, and use attribute `c` to store attributes associated with the line segment. We will use more of these features in the next chapter.

All the endpoints in `Segment` are based on the `Point` class that we have previously developed in Listing 2.1. We also define a suite of logic relations between two segments by overriding built-in Python functions such as `__eq__` (for equality) and `__lt__` (less than). Segment s_1 is said to be smaller than segment s_2 if s_1 is completely below s_2. We also include the function `contains` to test if a point is one of the endpoints of a segment.

Listing 2.6: Data structure for line segments (linesegment.py).

```
from point import *
from sideplr import *

## Two statuses of the left endpoint
ENDPOINT = 0 ## original left endpoint
INTERIOR = 1 ## interior in the segment

```

```python
class Segment:
    """
    A class for line segments.
    """
    def __init__(self, e, p0, p1, c=None):
        """
        Constructor of Segment class.
        Input
          e: segment ID, an integer
          p0, p1: endpoints of segment, Point objects
        """
        if p0>=p1:
            p0,p1 = p1,p0           # p0 is always left
        self.edge = e               # ID, in all edges
        self.lp = p0                # left point
        self.lp0 = p0               # original left point
        self.rp = p1                # right point
        self.status = ENDPOINT      # status of segment
        self.c = c                  # c: feature ID
    def __eq__(self, other):
        if isinstance(other, Segment):
            return (self.lp==other.lp and self.rp==other.rp)\
                or (self.lp==other.rp and self.rp==other.lp)
        return NotImplemented
    def __ne__(self, other):
        result = self.__eq__(other)
        if result is NotImplemented:
            return result
        return not result
    def __lt__(self, other):
        if isinstance(other, Segment):
            if self.lp and other.lp:
                lr = sideplr(self.lp, other.lp, other.rp)
                if lr == 0:
                    lrr = sideplr(self.rp, other.lp, other.rp)
                    if other.lp.x < other.rp.x:
                        return lrr > 0
                    else:
                        return lrr < 0
                else:
                    if other.lp.x > other.rp.x:
                        return lr < 0
                    else:
                        return lr > 0
        return NotImplemented
    def __gt__(self, other):
        result = self.__lt__(other)
        if result is NotImplemented:
            return result
        return not result
    def __repr__(self):
        return "{0}".format(self.edge)
```

```
60     def contains(self, p):
61         """
62         Returns True if segment has p as an endpoint
63         """
64         if self.lp == p:
65             return -1
66         elif self.rp == p:
67             return 1
68         else:
69             return 0
```

The code in Listing 2.7 shows an example of how to calculate the intersection point between two line segments. We first use function `test_intersect` to determine if two given line segments will intersect (line 60), which can be done by testing the sides of the segments using the `sideplr` function (e.g., line 41). If both endpoints of a segment are on the same side of the other segment, no intersection will occur. Otherwise, we use the equations introduced at the beginning of this section in function `getIntersectionPoint` to compute the actual intersection point. This function assumes the two input segments indeed intersect (hence it is necessary to test intersection first). The two segments in the test data (lines 57 and 58) have an intersection point at (1.5, 2.5).

Listing 2.7: Calculating the intersection between two line segments (intersection.py).

```
1   from linesegment import *
2   from sideplr import *
3
4   def getIntersectionPoint(s1, s2):
5       """
6       Calculates the intersection point of two line segments
7       s1 and s2. This function assumes s1 and s2 intersect.
8       Intersection must be tested before calling this function.
9       """
10      x1 = float(s1.lp0.x)
11      y1 = float(s1.lp0.y)
12      x2 = float(s1.rp.x)
13      y2 = float(s1.rp.y)
14      x3 = float(s2.lp0.x)
15      y3 = float(s2.lp0.y)
16      x4 = float(s2.rp.x)
17      y4 = float(s2.rp.y)
18      if s1.lp < s2.lp:
19          x1,x2,y1,y2,x3,x4,y3,y4=x3,x4,y3,y4,x1,x2,y1,y2
20      if x1 != x2:
21          alpha1 = (y2-y1)/(x2-x1)
22      if x3 != x4:
23          alpha2 = (y4-y3)/(x4-x3)
24      if x1 == x2: # s1 is vertical
25          y = alpha2*(x1-x3)+y3
26          return Point([x1, y])
```

```python
        if x3==x4: # s2 is vertical
            y = alpha1*(x3-x1)+y1
            return Point([x3, y])
        if alpha1 == alpha2: # parallel lines
            return None
        # need to calculate
        x = (alpha1*x1-alpha2*x3+y3-y1)/(alpha1-alpha2)
        y = alpha1*(x-x1) + y1
        return Point(x, y)

def test_intersect(s1, s2):
    if s1==None or s2==None:
        return False
    # testing: s2 endpoints on the same side of s1
    lsign = sideplr(s2.lp0, s1.lp0, s1.rp)
    rsign = sideplr(s2.rp, s1.lp0, s1.rp)
    if lsign*rsign > 0:
        return False
    # testing: s1 endpoints on the same side of s2
    lsign = sideplr(s1.lp0, s2.lp0, s2.rp)
    rsign = sideplr(s1.rp, s2.lp0, s2.rp)
    if lsign*rsign > 0:
        return False
    return True

if __name__ == "__main__":
    p1 = Point(1, 2)
    p2 = Point(3, 4)
    p3 = Point(2, 1)
    p4 = Point(1, 4)
    s1 = Segment(0, p1, p2)
    s2 = Segment(1, p3, p4)
    s3 = Segment(2, p1, p2)
    if test_intersect(s1, s2):
        print getIntersectionPoint(s1, s2)
        print s1==s2
        print s1==s3
```

2.7 Point-in-polygon operation

Identifying whether a point is inside a polygon is one of the most important operations in everyday GIS uses. For example, when we click on a point on a digital world map, we would expect to quickly retrieve the information related to the region that contains the point. For a simple (not self-intersecting) convex polygon, we can determine that a point is inside a polygon if it is on the same side of all the edges of that polygon. This approach seems to be computationally intensive because of the many multiplications, and it does not work for concave polygons (Figure 2.4).

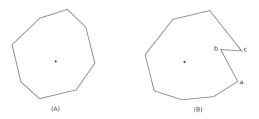

Figure 2.4 Point-in-polygon determination using the side of the point (dot) in relation to the line segments of a polygon, given a fixed sequence of points (clockwise or counterclockwise). (A) The point is always on the same side of all the segments. (B) The point is on the opposite sides of segments *ab* and *bc*

2.7.1 Even–odd algorithm

The even–odd algorithm is a popular method and is also known as the ray-casting or crossing number algorithm. This algorithm runs a crossing test that draws a half-line (or casts a ray) from the point. If the ray crosses the polygon boundary in an odd number of points, then we conclude that the point is inside the polygon. Otherwise, it is outside the polygon. Conveniently, we can draw the half-line horizontally (Figure 2.5). The overall process is straightforward but we hope to avoid the actual calculation of intersections because such calculation can be time-consuming, especially when we have to do it repeatedly.

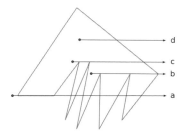

Figure 2.5 Point-in-polygon algorithm. The start point of half-line *a* is outside the polygon, and the start points of half-lines *b*, *c*, and *d* are inside the polygon

Figure 2.6 illustrates the different cases of whether it is necessary to compute the intersection. In this figure, we have a horizontal half-line that starts at point A, and a number of line segments that are labeled as *a* through *e* and *b'*. We try to test whether a line segment intersects the half-line without actually trying to compute the intersection point. Segments *a*, *d*, and *e* will not intersect the half-line because they have both X coordinates on one side of the origin of the half-line (*a*), or both Y coordinates on one side of the half-line (*d* and *e*). We count segment *c* as a crossing because we have an intersection, but we do not need to compute the intersection point because we know for sure that segment *c* crosses the half-line: the Y coordinates of segment *c* are on different sides of the half-line and both X coordinates are to the right of point A. For segments *b* and *b'*, we cannot conclude whether they intersect the half-line immediately and must compute the intersection point. Fortunately, we only need to compute the *X* coordinate of the intersection to decide whether the segment intersects the half-line or not. For segment *b*, since the *X* coordinate of the intersection point is not on the half-line itself but to the left of point *A* (i.e., the intersection point is on the dashed line), it does not intersect the half-line. For segment *b'*, since the intersection point is to the right of point *A*, we can conclude that it crosses the half-line.

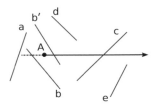

Figure 2.6 Cases for calculating intersection points

There is an important exception that represents multiple cases in real applications. What happens if the ray goes through the two endpoints of a line segment? What happens if the ray crosses through one of the endpoints? In these case, we can automatically add a very small value to the endpoints in question so that they are in theory "above" the ray. The consequence of this addition is that we make sure that a ray can only cross a line or not, and there is no third possibility. In the actual implementation of the algorithm, we treat an endpoint as "above" the half-line if its *Y* coordinate is "greater than" or "equal to" the *Y* coordinate of the half-line. Otherwise, it is treated as below the half-line. The half-line *b* in Figure 2.5, for example, will be considered as crossing the polygon boundary five times (instead of once), and the half-line *a* crosses the polygon eight times.

The function `pip_cross` in Listing 2.8 returns whether a point is inside a polygon using the even–odd algorithm. Here we dynamically maintain two points `p1` and `p2` that are the two consecutive vertices on the polygon boundary. We use variables `yflag1` and `yflag2` to indicate the vertical side of the two vertices with respect to the half-line (above/equal or below), and `xflag1` and `xflag2` to indicate the horizontal side (left/equal or right) of the two points with respect to the given point. In line 25 we check if the points `p1` and `p2` are on different sides of the half-line. If that is the case, we further check if they are on different horizontal sides of the point in line 28. If both vertices are on the right side of the point, we count one crossing. Otherwise, we will have to compute the *X* coordinate of the intersection (line 34) and decide if there is a crossing (line 35).

Listing 2.8: The even–odd algorithm for the point-in-polygon test (point_in_polygon.py).

```
1  import math
2  from point import *
3
4  def pip_cross(point, pgon):
5      """
6      Determines whether a point is in a polygon. Code adopted
7      from the C program in Graphics Gems IV by Haines (1994).
8      Input
```

```
        pgon: a list of points as the vertices for a polygon
        point: the point
      Output
        Returns a boolean value of True or False and the number
        of times the half-line crosses the polygon boundary
      """
      numvert = len(pgon)
      tx=point.x
      ty=point.y
      p1 = pgon[numvert-1]
      p2 = pgon[0]
      yflag1 = (p1.y >= ty)           # p1 on or above point
      crossing = 0
      inside_flag = 0
      for j in range(numvert-1):
          yflag2 = (p2.y >= ty)       # p2 on or above point
          if yflag1 != yflag2:        # both sides of half-line
              xflag1 = (p1.x >= tx)   # left-right side of p1
              xflag2 = (p2.x >= tx)   # left-right side of p2
              if xflag1 == xflag2:    # both points on right side
                  if xflag1:
                      crossing += 1
                      inside_flag = not inside_flag
              else:
                  m = p2.x - float((p2.y-ty))*\
                      (p1.x-p2.x)/(p1.y-p2.y)
                  if m >= tx:
                      crossing += 1
                      inside_flag = not inside_flag
          yflag1 = yflag2
          p1 = p2
          p2 = pgon[j+1]
      return inside_flag, crossing

if __name__ == "__main__":
    points = [ [0,10], [5,0], [10,10], [15,0], [20,10],
               [25,0], [30,20], [40,20], [45,0], [50,50],
               [40,40], [30,50], [25,20], [20,50], [15,10],
               [10,50], [8, 8], [4,50], [0,10] ]
    ppgon = [Point(p[0], p[1]) for p in points ]
    inout = lambda pip: "IN" if pip is True else "OUT"
    point = Point(10, 30)
    print "Point %s is %s"%(
        point, inout(pip_cross(point, ppgon)[0]))
    point = Point(10, 20)
    print "Point %s is %s"%(
        point, inout(pip_cross(point, ppgon)[0]))
    point = Point(20, 40)
    print "Point %s is %s"%(
        point, inout(pip_cross(point, ppgon)[0]))
    point = Point(5, 40)
    print "Point %s is %s"%(
        point, inout(pip_cross(point, ppgon)[0]))
```

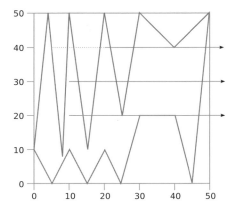

Figure 2.7 Test cases for the even–odd algorithm

We test the even–odd algorithm using the polygon and points illustrated in Figure 2.7. The program reports the following results:

```
Point (10,30) is IN
Point (10,20) is IN
Point (20,40) is IN
Point (5,40) is OUT
```

2.7.2 Winding number algorithm

In the two cases in Figure 2.4, we note that the segments of the polygon wind around the inside points. A polygon that does not contain a point does not wind around the point. The winding number algorithm uses this feature to test whether a specified point is inside or outside a polygon. In general, given two vectors in n dimensions $X = (x_1, x_2, \ldots, x_n)$ and $Y = (y_1, y_2, \ldots, y_n)$, we have $X \cdot Y = |X|\,|Y| \cos\theta$, where θ is the angle between the two vectors, $X \cdot Y$ is the dot product between the two vectors that equals $\sum_{i=1}^{n} x_i y_i$, and $|X|$ and $|Y|$ are the norms (or lengths) of X and Y, respectively. Given a line segment between points $v_i = (x_i, y_i)$ and $v_{i+1} = (x_{i+1}, y_{i+1})$, and a point $p = (x, y)$, the angle between p and the line segment is denoted by θ_i and can be computed as

$$\theta_i = \arccos\left[\frac{\overrightarrow{v_i p} \cdot \overrightarrow{v_{i+1} p}}{|\overrightarrow{v_i p}|\,|\overrightarrow{v_{i+1} p}|}\right]$$

$$= \arccos\left[\frac{(x-x_i)(x-x_{i+1}) + (y-y_i)(y-y_{i+1})}{\sqrt{(x-x_i)^2 + (y-y_i)^2}\,\sqrt{(x-x_{i+1})^2 + (y-y_{i+1})^2}}\right],$$

where $\overrightarrow{v_i p}$ is the vector from point v_i to p and $|\overrightarrow{v_i p}|$ is the length of the vector. Now we assume the polygon has n points and we have $v_1 = v_n$, indicating a closed polygon.

The sum of the angles between the point and all the n line segments of a polygon can therefore be computed as

$$wn = \frac{1}{2\pi} \sum_{i=1}^{n-1} \theta_i.$$

The point is inside the polygon if its winding number is not zero.

The winding number algorithm as presented above is not computationally efficient due to the use of trigonometric functions. This problem can be addressed by checking the winding of polygon line segments around a point by their directions. If a line goes upward crossing the point's Y coordinate, the winding number increases by 1, and if it goes downward, the winding number decreases by one. As demonstrated in Figure 2.8, the left point of both horizontal lines would be considered as inside the polygon using the winding number approach, as line a has one upward line and line b has two upward lines crossing. Essentially this approach becomes the same as the crossing method described before and they both have the same efficiency. The code for the two winding number methods is presented in Listing 2.9, where the function `pip_wn` is the conventional winding number algorithm, and `pip_wn1` is the crossing edge version. The test data used in the listing reflects a situation that is similar to that in Figure 2.8.

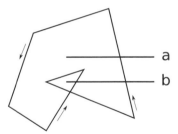

Figure 2.8 The winding number approach to point-in-polygon testing. The arrows are used to indicate the direction of each line segment of the polygon, using an overall counterclockwise order

Listing 2.9: The winding number algorithm (point_in_polygon_winding.py).

```
import math
from point import *

def is_left(p, p1, p2):
    """
    Tests if point p is to the left of a line segment
    between p1 and p2
    Output
        0 the point is on the line
        >0 p is to the left of the line
        <0 p is to the right of the line
```

```
        """
        return (p2.x-p1.x)*(p.y-p1.y) - (p.x-p1.x)*(p2.y-p1.y)

def pip_wn(pgon, point):
    """
    Determines whether a point is in a polygon using the
    winding number algorithm using trigonometric functions.
    Code adopted from the C program in Graphics Gems IV
    (Haines, 1994).
    Input
      pgon: a list of points as the vertices for a polygon
      point: the point
    Output
      Returns a boolean value of True or False and the number
      of times the half-line crosses the polygon boundary
    """
    if pgon[0] != pgon[-1]:
       pgon.append(pgon[0])
    n = len(pgon)
    xp = point.x
    yp = point.y
    wn = 0
    for i in range(n-1):
        xi = pgon[i].x
        yi = pgon[i].y
        xi1 = pgon[i+1].x
        yi1 = pgon[i+1].y
        thi = (xp-xi)*(xp-xi1) + (yp-yi)*(yp-yi1)
        norm = (math.sqrt((xp-xi)**2+(yp-yi)**2)
               * math.sqrt((xp-xi1)**2+(yp-yi1)**2))
        if thi != 0:
            thi = thi/norm
        thi = math.acos(thi)
        wn += thi
    wn /= 2*math.pi
    wn = int(wn)
    return wn is not 0, wn

def pip_wn1(pgon, point):
    """
    Determines whether a point is in a polygon using the
    winding number algorithm without trigonometric functions.
    Code adopted from the C program in Graphics Gems IV
    (Haines, 1994).
    Input
      pgon: a list of points as the vertices for a polygon
      point: the point
    Output
      Returns a boolean value of True or False and the number
      of times the half-line crosses the polygon boundary
    """
    wn = 0
```

```
64        n = len(pgon)
65        for i in range(n-1):
66            if pgon[i].y <= point.y:
67                if pgon[i+1].y > point.y:
68                    if is_left(point, pgon[i], pgon[i+1])>0:
69                        wn += 1
70            else:
71                if pgon[i+1].y <= point.y:
72                    if is_left(point, pgon[i], pgon[i+1])<0:
73                        wn -= 1
74        return wn is not 0, wn
75
76    if __name__ == "__main__":
77        pgon = [ [2,3], [7,4], [6,6], [4,2], [11,5],
78                 [5,11], [2,3] ]
79        point = Point(6, 4)
80        ppgon = [Point(p[0], p[1]) for p in pgon ]
81        print pip_wn(ppgon, point)
82        print pip_wn1(ppgon, point)
```

There is an obvious problem with the winding number algorithm: the start (left) point of half-line *b* in Figure 2.8 will be counted as inside the polygon because the winding number is greater than zero. The even–odd algorithm would report that the point is outside the polygon because the half-line intersects the polygon twice. There has been some debate about whether such a point should be considered to be inside or outside. However, we can clear the myth by forcing polygons to be simple, meaning the edges of a polygon do not intersect other edges of the polygon, except of course at the endpoints. If this assumption holds, we should have two polygons in the figure and therefore the smaller polygon contains the point but the larger one does not.

2.8 Map projections

Map projection refers to the process of transforming geographic locations from a three-dimensional spherical coordinate system of longitudes and latitudes into two-dimensional Cartesian space. Such a transformation is often necessary because in this way we can project the locations on the earth's surface onto a two-dimensional, flat sheet for greater convenience. The output of a map projection essentially transforms from the geographic coordinate system to a projected coordinate system, and there have been many different ways of doing so. Here, we only introduce two projection methods that represent two different kinds of transformation approach.

The Robinson projection (Figure 2.9) is one of the most commonly used projections in cartography and beyond. It is not designed to preserve area or local shapes, but instead to show the world so that the high latitude areas can be clearly displayed. The Robinson projection is different from many of the other projections because it is not entirely based on mathematical formulas. In general, its central meridian is a straight line that is 0.5072 as long as the equator. The parallels are straight lines and are equally spaced between

38°N and 38°S, but the spacing decreases toward the poles from these two latitudes. The poles in the Robinson projection are straight lines, each being 0.5322 as long as the equator. On each parallel, the meridians are equally spaced.

Figure 2.9 Robinson projection

More specifically, the origin of the coordinate system in the Robinson projection is at the intersection between the equator and the central meridian. The parameters in Table 2.1 were given by Robinson. For each of the 19 latitudes ranging from 0° to 90° in steps of 5°, Robinson gave the length of each parallel and the map distance from the parallel to the equator. Since the meridians on each parallel are equally spaced, as long as we know the length of a parallel, we can quickly compute where the longitude reading is on the parallel,

Table 2.1 Length and distance to equator of the parallels in the Robinson projection

Latitude (φ)	Length (A)	Distance to equator (B)
00	1.0000	0.0000
05	0.9986	0.0620
10	0.9954	0.1240
15	0.9900	0.1860
20	0.9822	0.2480
25	0.9730	0.3100
30	0.9600	0.3720
35	0.9427	0.4340
40	0.9216	0.4958
45	0.8962	0.5571
50	0.8679	0.6176
55	0.8350	0.6769
60	0.7986	0.7346
65	0.7597	0.7903
70	0.7186	0.8435
75	0.6732	0.8936
80	0.6213	0.9394
85	0.5722	0.9761
90	0.5322	1.0000

referenced to the central meridian, which is the X coordinate of the location. However, we only have the exact length of the given parallels (column 2 in Table 2.1). For the Y coordinate, we only know those at the given latitudes also (note that parallels are equally spaced between 38°N and 38°S). For any other latitude, we will have to interpolate the length (A) and distance to equator (B).

The interpolation, however, has been a bit of a guessing game because Robinson did not specify which interpolation method was used in his original work. Many researchers have come up with different ways to approximate the Robinson projection using different methods. A useful interpolation method is to use a polynomial function (curve) to fit the data (latitude and A or B) so that the curve passes through all the given points, meaning that, given one of the latitudes in Table 2.1, the function will yield exactly the same A or B value for that latitude.

Here we introduce Neville's algorithm to fit such a function. In general, we assume there is a polynomial $p(x)=\sum_{i=0}^{n} a_i x^i$ that fits the $n+1$ pairs of input data $\{(x_i, y_i)\}$ such that $p(x_i)=y_i$, $0 \leq i \leq n$. Table 2.2 shows how Neville's algorithm is used to compute the value at x given an input of five pairs (x_i, y_i) (the first two columns). We denote by $p_{i,j}(x)$ the interpolation of the value at x given with degree $j-i$ (i.e., the exponent of x in the actual formula). At iteration 0 (column 3), each value equals the corresponding y value: we have a zero-degree polynomial of $p_{i,i} = x_i^0 y_i = y_i$. Starting from iteration 1 (in column 4), each value in the column is a linear interpolation of the two values above and beneath in the previous column. For example,

$$p_{1,2}(x) = \frac{x_2 - x}{x_2 - x_1} p_{1,1}(x) - \frac{x_1 - x}{x_2 - x_1} p_{2,2}(x)$$
$$= \frac{x_2 - x}{x_2 - x_1} y_1 - \frac{x_1 - x}{x_2 - x_1} y_2.$$

In an iterated process, we compute all the first-degree polynomials (the ones in the third column) and then move on to computing the next column until there is only one value to compute, which is determined by the number of pairs in the input.

Table 2.2 Neville's algorithm

x_0	y_0	$p_{0,0}(x)$				
			$p_{0,1}(x)$			
x_1	y_1	$p_{1,1}(x)$		$p_{0,2}(x)$		
			$p_{1,2}(x)$		$p_{0,3}(x)$	
x_2	y_2	$p_{2,2}(x)$		$p_{1,3}(x)$		$p_{0,4}(x)$
			$p_{2,3}(x)$		$p_{1,4}(x)$	
x_3	y_3	$p_{3,3}(x)$		$p_{2,4}(x)$		
			$p_{3,4}(x)$			
x_4	y_4	$p_{4,4}(x)$				

In sum, we can write the formula for Neville's algorithm using the determinant of a square matrix as

$$p_{i,j}(x) = \frac{1}{x_j - x_i} \begin{vmatrix} p_{i,j-1}(x) & x_i - x \\ p_{i+1,j}(x) & x_j - x \end{vmatrix},$$

or equivalently as

$$p_{i,j}(x) = \frac{(x_j - x)p_{i,j-1}(x) + (x - x_i)p_{i+1,j}(x)}{x_j - x_i}.$$

We implement Neville's algorithm in Listing 2.10. In this code, we do not actually need a two-dimensional array to store p_{ij} values. If we examine these values in Table 2.2 carefully, we can see that at iteration k ($0 \leq k \leq n$), we only need to store up to $n - k$ values, which would make it perfect to use a list of size n (line 12). The trick is that once the value of p[i] is updated, the new value will not affect the calculation in the same iteration (line 20).

Listing 2.10: Neville's algorithm (neville.py).

```
1  def neville(datax, datay, x):
2      """
3      Finds an interpolated value using Neville's algorithm.
4      Input
5        datax: input x's in a list of size n
6        datay: input y's in a list of size n
7        x: the x value used for interpolation
8      Output
9        p[0]: the polynomial of degree n
10     """
11     n = len(datax)
12     p = n*[0]
13     for k in range(n):
14         for i in range(n-k):
15             if k == 0:
16                 p[i] = datay[i]
17             else:
18                 p[i] = ((x-datax[i+k])*p[i]+ \
19                         (datax[i]-x)*p[i+1])/ \
20                         (datax[i]-datax[i+k])
21     return p[0]
```

We are now ready to program the Robinson projection (Listing 2.11). We first store the values in Table 2.1 in three lists at the beginning of the code. A function called `transform1` is used to convert a longitude and latitude pair into the coordinate system of the Robinson projection. We use the central meridian of 0 for our code, and we will discuss this issue later when we discuss the Mollweide projection. The function `transform1` transforms a pair of coordinates in longitude and latitude into the coordinates in the Robinson projection. Since the latitudes in the table are positive, we take the absolute value of the latitude if it turns out negative (line 62). We will convert it back at the end if that is the case. Line 65 ensures the input latitude is within the 90° bounds. Line 67 finds the index of the largest latitude in the table that is smaller than or equal to the input latitude. This is done using a function called `find_le` that uses a binary search method to efficiently find the target latitude, thanks to the sorted values in the `latitudes` list. This function can be found in an official Python tutorial.[1]

After finding the index of the latitude from the table that is to the immediate left of (smaller than) the input latitude, we need to determine the known latitudes and *A* or *B* values for interpolation. Using all the values in the table for interpolation is simple but risky. This is because the degree of the interpolating polynomial is $n - 1$, with n being the number of input pairs. Using all the values in the table will yield a polynomial formula of degree 18 (i.e., one of the variables in the formula has an exponent of 18). This is very likely to be an over-fit of the data, meaning we will get exactly the same values at the latitude intervals, but not so for the values in between. Here, we generally pick two values from the left (smaller) side of the input latitude and two from the right (greater) for interpolation. For example, if the input latitude is 12.5, we will use latitudes of 5, 10, 15, and 20 for interpolation and the function `find_le` will return 2, the index of latitude 10. This becomes trickier when the value is at one end of the table, where we may not have enough indices on one side. For example, an input of 2.5 will force us to use only one latitude on the left or smaller side, meaning we only have three latitudes, 0, 5, and 10, for interpolation.

With the first left index found by `find_le`, we can find the other indices using a function called `get_interpolation_range`. This function also gives us the flexibility to specify the number of indices on both sides. By default, we use two values from both sides here. Line 68 gets the range of the latitudes that are used to interpolate the *B* value or *Y* coordinate. To interpolate the distance from a parallel to the equator, however, we must consider those parallels between 38°N and 38°S, which are equally spaced. In those cases, we can simply specify only one latitude from each side of the input latitude. As we discussed above, when only two input values are given, Neville's algorithm will return a linear interpolation, which is exactly what we need for interpolating the equally spaced parallels. This is done in line 72.

We call Neville's algorithm twice to interpolate the distance between the input parallel and the equator (line 70) and the length of the parallel (line 74). The rest of the code makes sure the requirements of the Robinson projection are satisfied so that the length of the central meridian is 0.5072 as long as the equator (line 75), and the meridians are equally spaced on each parallel (lines 76 and 77).

---

[1] `https://docs.python.org/2/library/bisect.html`

Listing 2.11: Transforming points to the Robinson projection (transform1.py).

```
import bisect
from neville import *
from numpy import fabs

latitudes=[0, 5, 10, 15, 20, 25, 30, 35, 40, 45, 50, 55, 60,
           65, 70, 75, 80, 85, 90]

# length of parallels at each latitude in latitudes
A=[1.0000, 0.9986, 0.9954, 0.9900, 0.9822, 0.9730, 0.9600,
   0.9427, 0.9216, 0.8962, 0.8679, 0.8350, 0.7986, 0.7597,
   0.7186, 0.6732, 0.6213, 0.5722, 0.5322]

# length from each parallel to the equator
# these values must be multiplied by 0.5072
B=[0.0000, 0.0620, 0.1240, 0.1860, 0.2480, 0.3100, 0.3720,
   0.4340, 0.4958, 0.5571, 0.6176, 0.6769, 0.7346, 0.7903,
   0.8435, 0.8936, 0.9394, 0.9761, 1.0000]

def find_le(a, x):
    """Finds rightmost value less than or equal to x"""
    i = bisect.bisect_right(a, x)
    if i:
        return i-1
    raise ValueError

def get_interpolation_range(sidelen, n, i):
    """
    Finds the range of indices for interpolation
    in Robinson projection
    Input
      sidelen: the number of items on both sides of i,
               including i in the left
      n: the total number of items
      i: the index of the largest item smaller than the value
    Output
      ileft: the left index of the value (inclusive)
      iright: the right index of the value (noninclusive)
    """
    if i<sidelen:
        ileft = max([0, i-sidelen+1])
    else:
        ileft = i-sidelen+1
    if i>=n-sidelen:
        iright = min(n, i+sidelen+1)
    else:
        iright = i+sidelen+1
    return ileft, iright
```

```python
def transform1(lon, lat):
    """
    Returns the transformation of lon and lat
    on the Robinson projection.
    Input
      lon: longitude
      lat: latitude
    Output
      x: x coordinate (origin at 0,0)
      y: y coordinate (origin at 0,0)
    """
    n = len(latitudes)
    south = False
    if lat<0:
        south = True
        lat = fabs(lat)
    if lat>90:
        return
    i = find_le(latitudes, lat)
    ileft, iright = get_interpolation_range(2, n, i)
    y = neville(latitudes[ileft:iright],
                B[ileft:iright], lat)
    if lat<=38:
        ileft, iright = get_interpolation_range(1, n, i)
    x = neville(latitudes[ileft:iright],
                A[ileft:iright], lat)
    y = 0.5072*y/2.0
    dx = x/360.0
    x = dx*lon
    if south:
        y = -1.0 * y
    return x, y, ileft, i, iright
```

We will use two kinds of data to test the algorithm for the Robinson projection. The code in Listing 2.12 helps us generate a set of points for this purpose. First, we want to draw the graticule or the network formed by lines of parallels and meridians. Each of these lines consists of a set of points. We use a simple list to store all that information where each item in the list is another list of three values: the line ID, and the longitude and latitude of the point. The points of each line must be stored sequentially to ensure smooth rendering. We use the two `for` loops starting at line 6 to create each of the parallels in steps of 10°, and the two `for` loops starting at line 11 for the meridians at the same interval. For each meridian, we take one point in every 10° latitude except for the latitudes beyond 80°N and 80°S, where we sample the points more densely (1° per point) to ensure smooth curves.

The second data set is the world coastlines on a scale of 1:110 million,[2] and we want to plot that on the graticule. The original coastline data are stored in a shapefile. We

---

[2] http://www.naturalearthdata.com/downloads/110m-physical-vectors/

use the OGR module to read the coordinates and convert them to the simple data structure we use here. The name of the shapefile is specified as the input of the function (line 3) and then used in line 24 for the OGR module to open the file. More details about the OGR module and other related topics can be found in Appendix B. While the process is going on, we also record the number of lines in the graticule (`numgraticule`) and the total number of lines (`numline`), which will be used later to render the projection map.

Listing 2.12: Preparing the world map data (worldmap.py).

```
from osgeo import ogr

def prep_projection_data(fname):
    points=[]
    linenum = 0
    for lat in range(-90, 91, 10):
        for lon in range(-180, 181, 10):
            points.append([linenum, lon, lat])
        linenum += 1

    for lon in range(-180, 181, 10):
        for lat in range(-90, -80, 1):
            points.append([linenum, lon, lat])
        for lat in range(-80, 80, 10):
            points.append([linenum, lon, lat])
        for lat in range(80, 91, 1):
            points.append([linenum, lon, lat])
        linenum += 1

    numgraticule = linenum

    driveName = "ESRI Shapefile"
    driver = ogr.GetDriverByName(driveName)
    vector = driver.Open(fname, 0)
    layer = vector.GetLayer(0)

    for i in range(layer.GetFeatureCount()):
        f = layer.GetFeature(i)
        geom = f.GetGeometryRef()
        for i in range(geom.GetPointCount()):
            p = geom.GetPoint(i)
            points.append([linenum, p[0], p[1]])
        linenum += 1

    numline = max([p[0] for p in points]) + 1

    return points, numgraticule, numline
```

Basic Geometric Operations

Our final step in understanding the Robinson projection is to transform the above data sets in longitudes and latitudes into coordinates on the Robinson projection. We do this in the code in Listing 2.13. Here we use a powerful Python visualization and plotting module called Matplotlib. Of course we can draw the map using any GIS package, but here we stick with open source tools and Python specifically. More information about this module is given in Appendix A.

We first get the data in their original format (line 7) and then transform them to the Robinson projection (line 11) with the same data structure of line ID, X and Y coordinates (line 12). We start to render the data by first getting the frame where everything will be plotted (line 24). In a `for` loop that goes through each line in the data (line 14), we set the color (line 15) so that the graticule is displayed in light grey and the world coastlines are drawn in a darker grey. Note that color in Python can be specified in different ways, and here we first use the verbal index of "lightgrey" and then an HTML color format specifying the intensity of red, green, and blue (in two 16-bit digits) in a string following the "#" sign. Lines 19 and 20 get the sequence of X and Y coordinates in the line, respectively, which are used to plot a two-dimensional line object in Matplotlib (line 21). Note that the lines are not displayed yet because they are just added into a graph that will be shown later using command `show`.

The rest of the code consists of some routines necessary to draw the map correctly. Line 23 is especially crucial because we want to make sure the two axes are correctly scaled since they do not have the same length in the projection. Line 24 gets the current axes and makes some changes so that we do not show the axes with the map. The final result of this process is shown in Figure 2.9, which is generated using the commented-out line 28 in the code that saves the rendering into an encapsulated Postscript figure for publication.

Listing 2.13: Testing Robinson projection using the world map data (test_projection.py).

```
from osgeo import ogr
import matplotlib.pyplot as plt
from transform1 import *
from worldmap import *

fname = '../data/ne_110m_coastline.shp'
pp, numgraticule, numline = prep_projection_data(fname)

points=[]
for p in pp:
    p1 = transform1(p[1], p[2])
    points.append([p[0], p1[0], p1[1]])

for i in range(numline):
    if i<numgraticule:
        col = 'lightgrey'
    else:
        col = '#5a5a5a'
    ptsx = [p[1] for p in points if p[0]==i]
    ptsy = [p[2] for p in points if p[0]==i]
    plt.plot(ptsx, ptsy, color=col)
```

```
23  plt.axis('scaled')
24  frame = plt.gca()
25  frame.axes.get_xaxis().set_visible(False)
26  frame.axes.get_yaxis().set_visible(False)
27  frame.set_frame_on(False)
28  #plt.savefig(robinson.eps',bbox_inches='tight',pad_inches=0)
29  frame.set_frame_on(True)
30  plt.show()
```

The Mollweide projection is an equivalent projection that preserves the area. This is a mathematical transformation based on the equations

$$x = \frac{\sqrt{8}}{\pi} R(\lambda - \lambda_0)\cos\theta,$$
$$y = \sqrt{2} R \sin\theta,$$

where R is the radius of the globe onto which the earth is projected (typically set to 1), λ is the longitude to be transformed, λ_0 is the longitude of the central meridian, and θ is a parametric angle that must satisfy the condition

$$2\theta + \sin 2\theta = \pi \sin\varphi,$$

where φ is the latitude to be transformed. The actual value of θ at each latitude φ must be estimated through an iterated algorithm such as the following Newton–Raphson method:

$$\Delta\theta' = -\frac{\theta' + \sin\theta' - \pi\sin\varphi}{1 + \cos\theta'}.$$

To use this iteration method, we start from the value of φ as the initial θ' to obtain the first $\Delta\theta'$. Then we set $\theta' + \Delta\theta'$ to be the new θ' and repeat the process. We continue until $\Delta\theta'$ decreases to a very small value that is close to zero. The final θ' value is then used to compute θ as

$$\theta = \theta'/2.$$

We implement this algorithm in function `opt_theta` in Listing 2.14, where the process stops when $\Delta\theta'$ reaches below 0.0000001. It should be noted that we normally express angles in degrees, but the Python trigonometric functions typically use radians as the unit; there is a need to ensure correct conversions between units. We use the

NumPy functions `degrees` and `radians` for this purpose (see Appendix A for an introduction to NumPy). Using the function on a latitude of −30°, for example, returns the following results:

```
>>> opt_theta(-30, True)
theta = -30.0
delta = -16.8015452415
theta = -46.8015452415
delta = -0.849241867829
theta = -47.6507871093
delta = -0.00275402735705
theta = -47.6535411367
delta = -2.92287909643e-08
-0.41585559653479859
```

Transforming coordinates to the Mollweide projection is simple, as implemented in the function `transform2` (line 25). In the input arguments, we include the longitude of the central meridian (`lon0`). This, however, requires a little more work because we need to make sure there are 180 degrees on both sides of the central meridian, while the input longitudes are from the original data. This is done in the code block from line 39.

Listing 2.14: Transforming points to the Mollweide projection (transform2.py).

```
from numpy import pi, cos, sin, radians, degrees, sqrt

def opt_theta(lat, verbose=False):
    """
    Finds optimal theta value using Newton-Raphson iteration.
    Input
      lat: the latitude value
      verbose: True to print intermediate output
    Output
      theta
    """
    lat1 = radians(lat)
    theta = lat1
    while True:
        dtheta = -(theta+sin(theta) -
                  pi*sin(lat1))/(1.0+cos(theta))
        if verbose:
            print "theta =", degrees(theta)
            print "delta =", degrees(dtheta)
        if int(1000000*dtheta) == 0:
            break
        theta = theta+dtheta
    return theta/2.0
```

```
def transform2(lon, lat, lon0=0, R=1.0):
    """
    Returns the transformation of lon and lat
    on the Mollweide projection.
    Input
      lon: longitude
      lat: latitude
      lon0: central meridian
      R: radius of the globe
    Output
      x: x coordinate (origin at 0,0)
      y: y coordinate (origin at 0,0)
    """
    lon1 = lon-lon0
    if lon0 <> 0:
        if lon1>180:
            lon1 = -((180+lon0)+(lon1-180))
        elif lon1<-180:
            lon1 = (180-lon0)-(lon1+180)
    theta = opt_theta(lat)
    x = sqrt(8.0)/pi*R*lon1*cos(theta)
    x = radians(x)
    y = sqrt(2.0)*R*sin(theta)
    return x, y
```

We can quickly test the Mollweide projection code by simply changing two small things in Listing 2.13: we make sure to import `transform2` and then change line 11 to `transform2`. Figure 2.10 shows the result of projecting the graticule and world coastlines onto the Mollweide projection.

Figure 2.10 Mollweide projection

An interesting feature of the Mollweide projection is that the longitudes that are 90° east and west of the central meridian are transformed to form a perfect circle. In this way,

Basic Geometric Operations 45

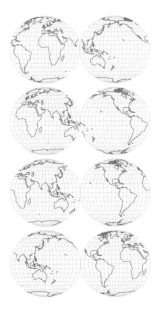

Figure 2.11 The world map projected onto two circles using the Mollweide projection. The central meridians in the left circle, from top to bottom, are 0°, 60°, 120°, and 180°, respectively

we can project the entire surface of the earth onto two circles and put them side by side. This feature has often been used in world atlases so that the eastern and western hemispheres can be nicely shown on one page. Figure 2.11 shows the two circles projected using a series of central meridians.

What happens if we do not modify the longitudes in the transform2 function in Listing 2.14? All the math will still work, but giving us an incorrect projection (Figure 2.12). In the exercises, we will see a problem in the data that makes it inappropriate to use central meridians other than the one at 0°. This problem, however, is with the data, not with the formula and the algorithm implementation we have discussed so far.

Figure 2.12 Mollweide projection with the central meridian at 90° without adjusting longitudes

2.9 Notes

The Haversine formula for calculating distance on the great circle can be found in Sinnott (1984). The point to line distance can be found in Deza and Deza (2010, p. 86).

Though many point-in-polygon algorithms exist (Haines, 1994; Huang and Shih, 1997), they are often designed based on the Jordan curve theorem (Jordan, 1887). The theorem specifies that a simple, non-self-intersecting polygon always divides the plane into exactly two parts, an interior region and an exterior region. The two parts share the curve as the boundary. In plain English, this is a simple polygon that does not intersect with itself. The concept of "point in polygon" or "inside a polygon" is intuitive in this case. However, the meaning of being inside a polygon is up for debate and the two algorithms, crossing number and winding number, may return different results.

The point-in-polygon algorithms introduced here all have a time complexity of $O(N)$ where N is the number of edges (Huang and Shih, 1997). While this appears light in terms of computing time, the total time used for point-in-polygon checking can increase quickly when a large number of polygons must be considered. To further reduce the time for all polygons, the indexing approaches described in a later chapter will be useful.

For a special polygon, such as monotonic, star, or convex, the point-in-polygon procedure can be simplified and therefore made more efficient. For example, for a convex polygon, the method can be achieved in $O(\log N)$ time because there can exist only two edges at most to intersect the half-line (O'Rourke, 1998). For convex (or monotonic in general) polygons, an algorithm can be specifically designed to speed up the test. For example, we can split the polygon into two sets of polylines, and in each set sort the lines according to their Y coordinates. In this way we can use a binary search algorithm to test whether an edge intersects with the half-line. W. Randolph Franklin[3] also proposed a test for convex polygons in four steps. First, find the equation of the infinite line that contains each edge. Second, express each equation as an expression $d = ax + by + c$. Points on the line will give zero. Third, standardize each equation so that if a point inside the polygon is substituted in, the result is positive, or equivalently an outside point will be negative. Fourth, now check the test point against every line. It is inside the polygon if and only if it is on the inside of every line.

Because Robinson's projection (Robinson, 1974) is not based on mathematical formulas,[4] many researchers have tried to replicate his original work (Richardson, 1989; Snyder, 1990; Ipbuker, 2004). It is also suggested that Robinson used the Aitken method

---

[3]PNPOLY – Point Inclusion in Polygon Test (`http://www.ecse.rpi.edu/Homepages/wrf/Research/Short_Notes/pnpoly.html`).

[4]Robinson himself also discussed this in a *New York Times* article (Wilford, 1988), in which he was quoted as saying: "I started with a kind of artistic approach. I visualized the best-looking shapes and sizes. I worked with the variables until it got to the point where, if I changed one of them, it didn't get any better. Then I figured out the mathematical formula to produce that effect. Most mapmakers start with the mathematics."

in his original work (Richardson, 1989), which is similar to Neville's algorithm used in this book, but considered to be obsolete (Press et al., 2002, p. 111). Mathematical details of many projections, including the Mollweide projection, can be found in Snyder (1987).

2.10 Exercises

1. Write a Python program to compute the Manhattan distance between two points.

2. In our discussion about polygon area calculation, we did not mention those polygons that may have holes. Will the equation we introduced here handle holes? If not, write a Python program that can be used to address this issue. A good place to look is the discussion in Appendix B.1.4 where we explore how to project and show complicated polygons.

3. Manually design a set of polygons and test the point-in-polygon algorithms we discussed in this chapter. Your polygons can be convex or concave, can have different shapes, and can be self-intersecting (or not).

4. Another way to test the polygons is to utilize something "real." By that we mean a real data set. Refer to the appendix on GDAL/OGR (Appendix B) to see how we can actually convert a real-world GIS data file into the simple data structure we used in this chapter, and test the point-in-polygon algorithms.

5. We mentioned that the angle version of the winding number algorithm is time-consuming because it relies on the use of trigonometric functions. Is this the case? Again, we may need to use the functions in GDAL/OGR that are discussed in Appendix B.

6. We have fixed the problem of the central meridian in the code for the Mollweide projection. However, if you really use a central meridian other than 0°, many of the coastlines will be stretched horizontally in the wrong direction. This is because our data are so tied to the graticule that ranges from −180° (east) to 180° (west). Also, if we use a central meridian that is not a multiple of 10°, we will have a pair of meridians projected in a different interval. Find a way to address this problem so we can correctly project the data using any central meridian.

7. The Python program for the Robinson projection does not include how to handle different central meridians. If you have successfully addressed Question 6, now would be the perfect time to continue and add a user-specified central meridian to the Robinson projection code.

8. Now that we have a good handle on the Robinson projection, we should be in a good position to test a few different projections. For example, can you use a similar approach (as in Table 2.1) to transforming the world map on to a triangular graticule where the south pole is a line of the same length as the central meridian and the north pole is a point? There are certainly more interesting shapes of the graticule to explore.

9. In most projections, the actual scale varies from point to point. Now that we know how to get the data for the graticule and know how to compute the distance in different coordinate systems (spherical or Cartesian), we can examine the scale factor on a map projection in a systematic manner. Write a Python program to do that and report your results on a map. What are the differences in scale distortion patterns in the Robinson and Mollweide projections?

10. We presented a series of maps projected in the Mollweide projection in Figure 2.11 without providing the code. The world coast data used in the figure is the same as the other maps and the `transform2` code in Listing 2.14 is used. Write a Python program to replicate the maps in this figure.

3

Polygon Overlay

And outside the walls, for half a mile around in every direction, the air was scented with the heavy rich smell of melting chocolate!

Roald Dahl, *Charlie and the Chocolate Factory*

It is a common task in GIS to calculate the intersections between line segments. For example, we may want to know where the path of a hurricane passes the coast of a country and the size of the area affected. This chapter contains fundamental algorithms that can be used to address such issues.

3.1 Line segment intersections

The algorithm for calculating the intersection between two line segments can be extended to the case of a set of line segments. A straightforward approach is simply to run a loop so that each pair of line segments is tested for intersection and therefore all possible intersection points are checked. This brute-force approach works, but it can be time-consuming, especially when the number of line segments is large and many line segments cannot intersect, as in the case of computing the intersection points of two sets of polygon maps, where line segments on one map will only intersect with a few lines on the other map.

Here, we introduce a more efficient algorithm that relies on a mechanism such that intersections are only considered when they are likely to occur. But where are those events for intersections to occur? Generally speaking, if we examine the segments locally, we can observe that only those segments that are close to each other have a chance of intersecting. For the four segments in Figure 3.1, lines S0 and S1 are more likely to intersect at locations e0, e1, e2, and e3 because their endpoints are getting closer at these locations than other segments (S0 and S2, or S1 and S2) are. Moving on toward the right of the figure, at locations after e3, the trend becomes different. For example, at location e4, segment S1 is closer to S2 and is more likely to intersect with S2 (indeed they intersect) instead of with S0. To take this line of thought further, let us imagine there

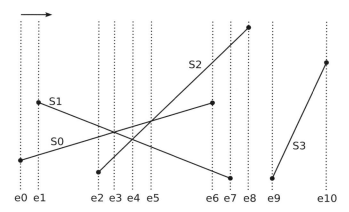

Figure 3.1 Line segment intersection using sweep lines

is a sweep line (dotted line) that moves from left to right to cross all the segments of interest. At any one time, we only need to consider the segments that are intersecting with the sweep line. More specifically, we only consider the events when the sweep line touches either the endpoints or the intersection points. Since the sweep line moves to the right, we will make sure that all the intersections to the left of the line are considered. In the case shown in Figure 3.1, the sweep line intersects S0 and S1, which means we will test if these two segments intersect. In this case, they do and we will make sure their intersection point is considered as an event when the sweep line moves there. It is important to note that other line segments to the right of the sweep line may intersect with S0 and S1, but we will visit those cases, if true, when the sweep line moves to those points.

We maintain a list of line segments that are related to an event point. If an event point is the left endpoint of a segment, we add that segment to the list. When we reach the right endpoint of a segment, the segment will then be removed from the list. At each event point, the line segments in the list are sorted in the same order as the segments "rise" from bottom to top. Therefore, for event e1 in Figure 3.1, the segments are listed as S0 and S1. When line segments intersect, a new interior event will be created. When the sweep line passes such an event point, it is clear that the order of the line segments meeting at the point will switch in the segment list. For example, the segment order at event point e3 is S2, S0, and S1. Since the event point is an interior point between S1 and S0, at the next event point e4, the order becomes S2, S1, and S0, which is the order of segments if we count from the bottom between events e3 and e4. Why is it important to maintain the correct order? Because the order dictates how intersection testing can be conducted. For example, at event point e1, we only need to check if S1 intersects with the closest (adjacent) segment in the list, which in this case is S0. The same logic applies to interior points, such as e3, where we test whether segments meeting at the point (i.e., S0 and S1) intersect with their adjacent segments (i.e., S2).

The events are maintained as a list that is sorted so that the events on the left are always placed in front of those on the right. To organize the list of segments efficiently at each event point, we use a balanced tree where each node is a segment. The left branch of a node is a segment that is below the segment represented by the node. Similarly, the segment represented by the right branch of a node is above the segment of the node. Using the example line segments on the left of Figure 3.2, the balanced binary tree for the event indicated by the dotted line is illustrated on the right.

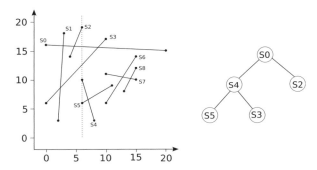

Figure 3.2 Sweep line. Left: a sample set of line segments where the dotted line indicates an event. Right: a balanced binary tree to store line segments at the event point

The processes discussed above are formally presented in the Bentley–Ottmann algorithm, also known as the sweep line algorithm. Here we first use pseudo-code to describe the process (Listing 3.1) and then we will discuss the implementation of the algorithm in Python.

Listing 3.1: The Bentley–Ottmann algorithm for testing intersections among line segments.

```
1   Initialize event queue EQ that includes the endpoints of
    all segments. Each left endpoint should also be associated
    with the line segment ID. After initialization, EQ will be
    sorted based on the X coordinate of each endpoint.
2   Initialize an empty binary tree, T, to store line segments
    at each event point.
3   While EQ is not empty
4      e = the next event in EQ and delete it from the queue
5      L = set of segments that have e as their left endpoint
6      R = the set of segments with e as their right endpoint
7      I = the set of segments with e as their interior point
8      If the union of L, R, and I has more than 1 segment
9         Report intersection at e
10     Delete segments in R and I from T
11     Insert segments in L and I into T
12     If both L and I are empty
```

```
13            s = the first segment in R and sr and sl are the
                      right and left branches of s in T
14            If sr and sl intersect at a point not in EQ
15                Add the intersection point into EQ
16        Else
17            s' = leftmost segment of L and I in T
18            s'' = rightmost segment of L and I in T
19            sl = the left node of s' in T
20            sr = the right node of s'' in T
21            If s' and sl intersect at a point not in EQ
22                Add the intersection point into EQ
23            If s'' and sr intersect at a point not in EQ
24                Add the intersection point into EQ
```

To implement the sweep line algorithm, we will need to develop some data structures to store and manage the events and the event queue (Listing 3.2). The Event class contains the point and its associated segments. The segments, however, will only be stored if the event is a left endpoint, as suggested in line 1 of the pseudo-code of the algorithm. The EventQueue class is initialized using a list of line segment objects. Note that only unique points are listed in the queue and the segments associated with the left endpoint will be added to the edges of the event (line 35). Additional functions are provided in the EventQueue class to help facilitate the operations. Specifically, the add function inserts a new event into the sorted queue. The find function returns the index of first appearance when the event or a point is found in the queue; it returns −1 if the point is not in the queue.

Listing 3.2: Event queue module (line_seg_eventqueue.py).

```python
from point import *

class Event:
    """
    An event in the sweep line algorithm. Each Event object
    stores the event point and the edges associated with the
    point.
    """
    def __init__(self, p=None):
        self.edges = []       # edges associated with the event
        self.p = p            # event point
    def __repr__(self):
        return "{0}{1}".format(self.p, self.edges)

class EventQueue:
    """
    An event queue in the sweep line algorithm.
    """
    def __init__(self, lset): # initialize the event queue
        """
        Constructor of EventQueue.
```

```
           Input
             lset: a list of Segment objects. The left point of
                   each segment is used to create an event
           Output
             A sorted list of events as a member of this class
           """
           if lset == None:
               return
           self.events = []
           for l in lset:
               e0 = Event(l.lp)
               inx = self.find(e0)
               if inx == -1:
                   e0.edges.append(l)
                   self.events.append(e0)
               else:
                   self.events[inx].edges.append(l)
               e1 = Event(l.rp)
               if self.find(e1) == -1:
                   self.events.append(e1)
           self.events.sort(key=lambda e: e.p)

       def add(self, e):
           """
           Adds event e to the queue, updates the list of events
           """
           self.events.append(e)
           self.events.sort(key=lambda e: e.p)

       def find(self, t):
           """
           Returns the index of event t if it is in the queue.
           Otherwise, returns -1.
           """
           if isinstance(t, Event):
               p = t.p
           elif isinstance(t, Point):
               p = t
           else: return -1
           for e in self.events:
               if p == e.p:
                   return self.events.index(e)
           return -1

       def is_empty(self):
           return len(self.events) == 0
```

When implementing the algorithm, because of the frequent operations on the line segments at each event point, it is more convenient and efficient to maintain the segments in a tree structure instead of just using a list data structure. Any balanced binary tree (such as an AVL[1] or a red–black tree) will be sufficient as long as it can support common operations such as insertion, deletion, and converting a tree to a sorted list. In our implementation, we simply utilize the AVL tree in a Python module called bintrees.[2]

In the implementation of the Bentley–Ottmann algorithm (Listing 3.3), we need a few functions to help retrieve some important information from the AVL tree. First, the function `get_edges` returns all the edges in a tree that contain a point as their right endpoint or as the interior point. Here, an interior point comes from two cases: when the point is already an intersection that has been calculated or when it is a left endpoint but is on another segment. The function `get_lr` returns the left and right neighbors of a segment in the tree; when the left or right neighbor does not exist, it returns the segment itself. Finally, the function `get_lrmost` returns the left- and rightmost segments of a list in the tree.

Listing 3.3: Implementation of the Bentley–Ottmann algorithm for line segment intersection (line_seg_intersection.py).

```python
from bintrees import AVLTree
from point import *
from intersection import *
from line_seg_eventqueue import *

def get_edges(t, p):
    """
    Gets the edges that contain point p as their right
    endpoint or in the interior
    """
    lr = []
    lc = []
    for s in AVLTree(t):
        if s.rp == p:
            lr.append(s)
        elif s.lp == p and s.status == INTERIOR:
            lc.append(s)
        elif sideplr(p, s.lp, s.rp) == 0:
            lc.append(s)
    return lr, lc

def get_lr(T, s):
    """
    Returns the left and right neighbors (branches) of p in T
    """
    try:
        sl = T.floor_key(s)
```

---

[1] AVL stands for Adelson-Velskii and Landis, the inventors of the self-balancing binary tree where the height difference between any two subtrees is at most 1.

[2] https://pypi.python.org/pypi/bintrees/2.0.1

```
28            except KeyError:
29                sl = None
30            try:
31                sr = T.ceiling_key(s)
32            except KeyError:
33                sr = None
34            return sl, sr
35
36     def get_lrmost(T, segs):
37         """
38         Finds the leftmost and rightmost segments of segs in T
39         """
40         l = []
41         for s in list(T):
42             if s in segs:
43                 l.append(s)
44         if len(l) < 1:
45             return None, None
46         return l[0], l[-1]
47
48     def find_new_event(s1, s2, p, q):
49         ip = intersectx(s1, s2)
50         if ip is None:
51             return False
52         if q.find(ip) is not -1:
53             return False
54         if ip.x>p.x or (ip.x==p.x and ip.y >= p.y):
55             e0 = Event()
56             e0.p = ip
57             e0.edges = [s1, s2]
58             q.add(e0)
59         return True
60
61     def intersectx(s1, s2):
62         """
63         Tests intersection of 2 segments and returns the
64         intersection point
65         """
66         if not test_intersect(s1, s2):
67             return None
68         p = getIntersectionPoint(s1, s2) # an intersection
69         return p
70
71     def intersections(psegs):
72         """
73         Implementation of the Bentley-Ottmann algorithm.
74         Input
75            psegs: a list of segments
76         Output
```

```
 77          intpoints: a list of intersection points
 78      """
 79      eq = EventQueue(psegs)
 80      intpoints = []
 81      T = AVLTree()
 82      L=[]
 83      while not eq.is_empty():      # for all events
 84          e = eq.events.pop(0)      # remove the event
 85          p = e.p                   # get event point
 86          L = e.edges               # segments with p as left end
 87          R,C = get_edges(T, p)     # p: right (R) and interior (C)
 88          if len(L+R+C) > 1:        # Intersection at p among L+R+C
 89              for s in L+R+C:
 90                  if not s.contains(p):   # if p is interior
 91                      s.lp = p            # change lp and
 92                      s.status = INTERIOR # status
 93              intpoints.append(p)
 94              R,C = get_edges(T, p)
 95          for s in R+C:
 96              T.discard(s)
 97          for s in L+C:
 98              T.insert(s, str(s))
 99          if len(L+C) == 0:
100              s = R[0]
101              if s is not None:
102                  sl, sr = get_lr(T, s)
103                  find_new_event(sl, sr, p, eq)
104          else:
105              sp, spp = get_lrmost(T, L+C)
106              try:
107                  sl = T.prev_key(sp)
108              except KeyError:          # only on last key
109                  sl = None
110              try:
111                  sr = T.succ_key(spp)
112              except KeyError:          # only on last key
113                  sr = None
114              find_new_event(sl, sp, p, eq)
115              find_new_event(sr, spp, p, eq)
116      return intpoints
```

We use the line segments illustrated in Figure 3.2 to demonstrate the algorithm (Listing 3.4).

Listing 3.4: Testing the Bentley–Ottmann algorithm (test_line_seg_intersection.py).

```
1  from line_seg_intersection import *
2
3  s = [ [[20,15],[0,16]], [[3,18],[2,3]], [[4,14],[6,19]],
4        [[10,17],[0,6]], [[8,3],[5,10]], [[6,6],[11,9]],
5        [[16,14],[10,6]], [[16,10],[10,11]],
```

```
 6              [[14,8],[16,12]] ]
 7
 8    psegs = [Segment(i, Point(s[i][0][0], s[i][0][1]),
 9                        Point(s[i][1][0], s[i][1][1]))
10             for i in range(len(s))]
11
12    ints = intersections(psegs)
13    print "There are", len(ints), "intersection points:"
14    print ints
```

Running the test program will result in the following output:

```
There are 7 intersection points:
[(2.4, 8.6), (2.9, 15.9), (4.7, 15.8), (6.6, 6.3),
    (8.7, 15.6), (13.3, 10.4), (15.1, 10.2)]
```

Table 3.1 shows the progression of the event queue and sorted segments in the tree, along with the report of each intersection point. To better represent the event points, we use L and R prefixes to indicate the left and right endpoints, respectively, and X to indicate interior or intersection points.

Table 3.1 Sweep line algorithm output

Event	Event queue	Segment in tree	Intersection
—	L3,L0,L1,R1,L2,L4,L5,R2,R4,L6,L7,R3,R5,L8,R7,R8,R6,R0	[]	
L3	L0,L1,R1,L2,L4,L5,R2,R4,L6,L7,R3,R5,L8,R7,R8,R6,R0	[3]	
L0	L1,R1,L2,L4,L5,R2,R4,X3X0,L6,L7,R3,R5,L8,R7,R8,R6,R0	[3, 0]	
L1	X3X1,R1,L2,L4,L5,R2,R4,X3X0,L6,L7,R7,R5,L8,R7,R8,R6,R0	[1, 3, 0]	X3X1
X3X1	X0X1,R1,L2,L4,L5,R2,R4,X3X0,L6,L7,R7,R5,L8,R7,R8,R6,R0	[3, 1, 0]	X0X1
X0X1	R1,L2,L4,L5,R2,R4,X3X0,L6,L7,R7,R5,L8,R7,R8,R6,R0	[3, 0, 1]	
R1	L2,L4,L5,R2,R4,X3X0,L6,L7,R7,R5,L8,R7,R8,R6,R0	[3, 0]	
L2	X0X2,L4,L5,R2,R4,X3X0,L6,L7,R7,R5,L8,R7,R8,R6,R0	[3, 2, 0]	X0X2
X0X2	L4,L5,R2,R4,X3X0,L6,L7,R7,R5,L8,R7,R8,R6,R0	[3, 0, 2]	
L4	L5,R2,R4,X3X0,L6,L7,R7,R5,L8,R7,R8,R6,R0	[4, 3, 0, 2]	
L5	R2,X4X5,R4,X3X0,L6,L7,R7,R5,L8,R7,R8,R6,R0	[5, 4, 3, 0, 2]	
R2	X4X5,R4,X3X0,L6,L7,R7,R5,L8,R7,R8,R6,R0	[5, 4, 3, 0]	X4X5
X4X5	R4,X3X0,L6,L7,R7,R5,L8,R7,R8,R6,R0	[4, 5, 3, 0]	
R4	X3X0,L6,L7,R7,R5,L8,R7,R8,R6,R0	[5, 3, 0]	X3X0
X3X0	L6,L7,R7,R5,L8,R7,R8,R6,R0	[5, 0, 3]	
L6	L7,R7,R5,L8,R7,R8,R6,R0	[6, 5, 0, 3]	
L7	R7,R5,L8,R7,R8,R6,R0	[6, 5, 7, 0, 3]	
R7	R5,L8,R7,R8,R6,R0	[6, 5, 7, 0]	
R5	X6X7,L8,R7,R8,R6,R0	[6, 7, 0]	X6X7
X6X7	L8,R7,R8,R6,R0	[7, 6, 0]	
L8	X7X8,R7,R8,R6,R0	[8, 7, 6, 0]	X7X8
X7X8	R7,R8,R6,R0	[7, 8, 6, 0]	
R7	R8,R6,R0	[8, 6, 0]	
R8	R6,R0	[6, 0]	
R6	R0	[0]	
R0	—	[]	

3.2 Overlay

The Bentley–Ottmann algorithm introduced above is efficient in calculating the intersection points given a set of line segments. However, this algorithm does not recognize segments from different maps and therefore does not treat them differently. In many GIS applications, we often need to compute the intersections of two sets of line segments, each from a specific map. This process is called map overlay and is arguably the most commonly used operation in GIS routines. For example, we have a map of population for each county in a state and another map that shows the distribution of forests in the state. Overlaying these two maps together can help understand such matters as how much forest is included in each county.

A critical task in overlaying two maps is to sort out the resulting polygons and make sure each final polygon correctly stores information from the two original maps. Different from simply obtaining the intersection points, this time we will need to deal with lines that bind the polygons and the polygons themselves. To better organize the procedure, we use a special data structure called the doubly connected edge list (DCEL) that accounts for the connections between the three geometric features of points, lines, and polygons. Let us consider the set of polygons illustrated in Figure 3.3. In a DCEL, each line segment is represented using two half-edges, each pointing in opposite directions. The half-edge from vertex i to j is denoted by $e_{i,j}$. Given a half-edge such as $e_{1,6}$, its other half ($e_{6,1}$ in this case) is called its twin. Each polygon is called a face in DCEL. Each face has a set of half-edges that defines its outline and these edges form a cycle. We will make sure that each face is always to the left of the directed half-edge. For example, face f_2 in the figure has an outline defined by the half-edges $e_{2,1}$, $e_{1,6}$, $e_{6,7}$, $e_{7,3}$, and $e_{3,2}$. Further, each half-edge has an origin point and we specify the `prev` half-edge as the half-edge that points to the origin on the boundary of the same face, and, by the same logic, the `next` half-edge as the half-edge pointed to by this half-edge on the boundary.

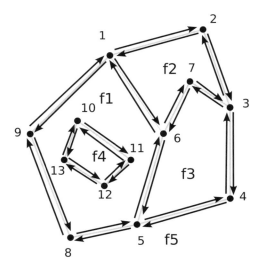

Figure 3.3 An example of a doubly connected edge list

We need three data structures to store the complete information for a set of polygons using a DCEL. First, we store the index of each vertex and all the half-edges originating from it (Table 3.2). Second, for each half-edge, we store its index, origin, twin, the face on its left (incident face), along with the `prev` and `next` half-edges as shown in Table 3.3, where only some of the half-edges are listed. Finally, for each face (Table 3.4), we use its index, one of the half-edges that form the outline of the face (outer component), and the half-edges inside the face (inner components). The inner components of a face are crucial to record the holes in a polygon. Each hole will require a half-edge. When a polygon has multiple holes, we will need to have multiple half-edges. In our example, face f1 has one hole and therefore only needs one half-edge as its inner component (e.g., $e_{10,11}$). Face f_5 represents the outside of all the polygons and it does not have an outer component.

Table 3.2 Vertices of a doubly connected edge list

Vertex	Edge list
1	e1,2, e1,6, e1,9
2	e2,1, e2,3
3	e3,2, e3,7, e3,4
4	e4,3, e4,5
5	e5,6, e5,8, e5,4
6	e6,1, e6,7, e6,5
7	e7,3, e7,6
8	e8,5, e8,9
9	e9,1, e9,8
10	e10,11, e10,13
11	e11,10, e11,12
12	e12,13, e12,11
13	e13,10, e13,12

Table 3.3 Edges of a doubly connected edge list

Half-edge	Origin	Twin	Incident face	Next	Prev
e1,2	1	e2,1	f5	e2,3	e9,1
e1,6	1	e6,1	f2	e6,7	e2,1
e1,9	1	e9,1	f1	e9,8	e6,1
e2,1	2	e1,2	f2	e1,6	e3,2
e2,3	2	e3,2	f1	e3,4	e1,2
e3,2	3	e2,3	f2	e2,1	e7,3
e3,7	3	e7,3	f3	e7,6	e4,3
e3,4	3	e4,3	f5	e4,5	e2,3
...					
e6,1	6	e1,6	f1	e1,9	e5,6
...					
e9,1	9	e1,9	f5	e1,2	e8,9
...					
e13,12	13	e12,13	f4	e12,11	e10,13

Table 3.4 Faces of a doubly connected edge list

Face	Outer component	Inner components
f1	e1,9	e10,11
f2	e2,1	None
f3	e7,6	None
f4	e11,10	None
f5	None	e1,2

We implement the polygon overlay algorithm in Listing 3.5, where a function called `overlay` is used to return a list of intersection points between two polygon maps. This function takes two sets of input parameters: a list of segments (`psegs`) and an incomplete DCEL (`D`) that only contains information for vertices and half-edges that are directly copied from the two source maps (we describe below how to prepare these data). The main body of this function is unsurprisingly similar to the `intersections` function in Listing 3.3. The additional features here rely on testing whether an intersection point is between two maps. We make sure segments from each map get the same value for a unique `c` attribute that is different from the other maps. We can tell this easily by checking if the intersecting segments involve two different `c` values (line 88). If no such crossing happens then there is no need to update anything in `D`. When a crossing between two maps exists, there are three cases: the edge of one map passes the vertex of another, the edge of one map crosses the edge of another, or the vertex of one map lies on the vertex of another. Given the way polygons and line segments are represented, there are no other cases that involve both maps at the intersection points. In this book, we only show how the first case can be handled (line 93), while the other two cases can be treated in a similar fashion.

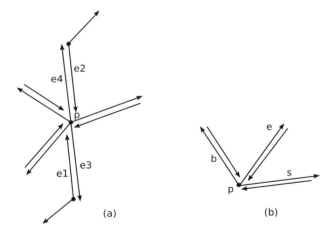

Figure 3.4 Handling the case where an edge passes through a vertex of another map

Handling the case where an edge of one map passes through the vertex (or vertices) of another map is processed in the function called `handle_edge_vertex` in Listing 3.5. While this case does not require the creation of new vertices, we will need to update the information for existing vertices, half-edges, and faces. Here, we first update information regarding vertices and half-edges (note that we use "hedge" to mean "half-edge" in the code). Faces will be updated later. Let p be the intersection point that is a vertex on one map, and e1 and e2 be the two half-edges of the original edge on the other map that passes through p (line 30). As shown in Figure 3.4a, we can keep the original origins of these two half-edges but need to change their `next` and `prev` half-edges to reflect the insertion of an intersection point that splits the original half-edges of e1 and e2. We will then create two new half-edges, e3 and e4, with p as their origin (lines 31 and 32). We will need to ensure e3 and e4 have correct `next` and `prev` half-edges (lines 34–37). Around the endpoints of the original edge, we have two pairs of twins: e1 and e3, and e2 and e4. We also know the next half-edge of e3 and e4 is the next half-edge of the previous e2 and e1, respectively. To update a new half-edge (denoted by e) around the intersection point p, we sort the edges with p as the origin using their angles and determine the two half-edges that are to the left (b) and right (s) of e (Figure 3.4b). The function `update_intersect_dcel` (line 6) details the processes needed to ensure the correct relations in DCEL regarding these half-edges. After the updates (lines 44 and 45), the two new half-edges will be added into D.

Listing 3.5: Overlay of two maps (overlay.py).

```
1   from line_seg_intersection import *
2   from dcel import *
3
4   class OvearlayError(Exception): pass
5
6   def update_intersect_dcel(hl, e):
7       """
8       Updates the hedges related to e given a hedge list hl
9       """
10      l = len(hl)
11      if l<2:
12          raise OvearlayError(
13              "Overlay/DCEL error: single edge for vertex")
14      big, small = l-1, 0
15      for i in range(l):
16          if e.angle > hl[i].angle:
17              big,small = i-1,i
18              break
19      b,s = hl[big], hl[small]
20      e.prevhedge = b.twin
21      e.twin.nexthedge = s
22      b.prevhedge = e.twin
```

```
23          b.twin.nexthedge = e
24          s.prevhedge = e.twin
25
26      def handle_edge_vertex(L, R, C, p, D):
27          e = C[0].edge
28          v1 = Vertex(C[0].lp0.x ,C[0].lp0.y)    # get the hedges for e
29          v2 = Vertex(C[0].rp.x ,C[0].rp.y)
30          e1,e2 = D.findhedges(v1,v2)            # origins:ends of e
31          e3 = Hedge(v1, Vertex(p.x,p.y))        # hedge,p as origin
32          e4 = Hedge(v2 ,Vertex(p.x,p.y))        # hedge,p as origin
33          # updating around endpoints of e:
34          e1.twin = e3
35          e3.twin = e1
36          e2.twin = e4
37          e4.twin = e2
38          e3.nexthedge = e2.nexthedge
39          e4.nexthedge = e1.nexthedge
40
41          v = D.findvertex(p)                    # updating around p
42          v.sortincident()
43          hl = v.hedgelist
44          update_intersect_dcel(v.hedgelist, e3)
45          update_intersect_dcel(v.hedgelist, e4)
46          # add two new hedges with p as origin
47          v.hedgelist.append(e3)
48          v.hedgelist.append(e4)
49          # e1, e2 updated in D due to references
50          D.hedges.append(e3)
51          D.hedges.append(e4)
52
53      def handle_edge_edge(L, R, C, p, D):
54          Pass
55
56      def handle_vertex_vertex(L, R, C, p, D):
57          Pass
58
59      def overlay(psegs, D):
60          """
61          Overlays polygons from two maps.
62          Input
63            psegs: a list of Segments. The c attribute in each
64                   segment indicates the source map
65            D: a partial DCEL with the original hedges and vertices
66          Output
67            intpoints: list of intersection points
68          """
69          eq = EventQueue(psegs)
70          intpoints = []
71          T = AVLTree()
```

```python
        L=[]
        while not eq.is_empty():      # for all events
            e = eq.events.pop(0)      # remove the event
            p = e.p                   # event point
            L = e.edges               # segments with p as left end
            R,C = get_edges(T, p)     # Intersection at p among L+R+C
            if len(L+R+C) > 1:
                for s in L+R+C:
                    if not s.contains(p):
                        s.lp = p
                        s.status = INTERIOR
                intpoints.append(p)
                R,C = get_edges(T, p)
                c1 = (L+R+C)[0].c
                cross = False
                for l in L+R+C:
                    if c1 is not l.c:
                        cross = True
                        break
                # Update its vertices and edge lists
                if cross is True:
                    if len(C) == 1:   # CASE 1: edge passes vertex
                        handle_edge_vertex(L, R, C, p, D)
                    if len(C) > 1:    # CASE 2: edge crosses edge
                        handle_edge_edge(L, R, C, p, D)
                    if len(C) == 0:   # CASE 3: vertex on vertex
                        handle_vertex_vertex(L, R, C, p, D)
            for s in R+C:
                T.discard(s)
            for s in L+C:
                T.insert(s, 1)
            if len(L+C) == 0:
                s = R[0]
                if s is not None:
                    sl, sr = get_lr(T, s)
                    y = find_new_event(sl, sr, p, eq)
            else:
                lp, lpp = get_lrmost(T, L+C)
                try:
                    sl = T.prev_key(lp)
                except KeyError:                # only on last key
                    sl = None
                try:
                    sr = T.succ_key(lpp)
                except KeyError:                # only on last key
                    sr = None
                find_new_event(sl, lp, p, eq)
                find_new_event(sr, lpp, p, eq)
        return intpoints
```

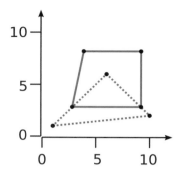

Figure 3.5 Testing the overlay algorithm

We use the example data shown in Figure 3.5 to demonstrate how the Bentley–Ottmann algorithm can be extended to compute the overlay between two maps. In this example, we assume that each map has a polygon, shown in solid and dotted lines, respectively. The code in Listing 3.6 sets up the initial steps and computes the overlay. The vertices in the two maps are represented in variables `pgon1` and `pgon2`, and their edges in `edges1` and `edges2`, respectively. These two maps are stored in a DCEL list as `d[0]` and `d[1]`. We extend an existing Python module to implement the DCEL data structure.[3] The `load` function of the DCEL module requires both lists of points and edges to create the DCEL object. We use another DCEL object `D` to store the data after overlay. We initialize `D` using the unions of vertices and half-edges from the two sources (lines 13 and 14). To overlay the maps, we still need to count the line segments from the two maps, which are stored in the line segment lists called `ps1` and `ps2`, where the member `c` is assigned 0 or 1 to distinguish the two maps (lines 28 and 31). These segments are then combined into one list called `s` (line 32) that is used as one of the two input variables of function `overlay`. The other input taken by the function is `D` and the function will update the information in `D`.

After running the `overlay` function, we should have all the correct relations among the intersecting vertices and related half-edges in `D`. Now it is time to make sure we have the faces registered correctly. Instead of reusing the faces in the original maps, faces in `D` will all be created from scratch. The code within the `while` loop in line 38 represents the cycling of line segments for each face. The first task is to know how many faces we will overlay and the half-edges that bound those faces. With a correct set of half-edges and vertices, given any half-edge, we can easily form a cycle by following the next half-edge stored in DCEL (line 43).

After checking the cycles, all the half-edges in `D` will be removed. But this is not a problem because we still have the half-edges, stored in variable `cycles`. The next job is to assign correct faces to each half-edge (line 57). We can easily test if a boundary cycle is the outer component of a face by checking the sign of the area. As discussed in the previous chapter, we will have a negative area value when traversing the edges in a counterclockwise order.

---

[3]`https://pypi.python.org/pypi/dcel/0.1.1`. The modified code is available on the Github site of this book.

Listing 3.6: Testing the overlay algorithm (test_overlay_1.py).

```python
from overlay import *

pgon1 = [ [3,3], [9,3], [8,8], [4,8] ]
edges1 = [ [0,1], [1,2], [2,3], [3,0] ]
pgon2 = [ [1,1], [6,6], [10,2] ]
edges2 = [ [0,1], [1,2], [2,0] ]

d=[ Dcel(), Dcel() ]
d[0].load(pgon1, edges1)
d[1].load(pgon2, edges2)

D = Dcel()
D.vertices = d[0].vertices + d[1].vertices
D.hedges = d[0].hedges + d[1].hedges

s1 = []
for i in range(len(pgon1)-1):
    s1.append([pgon1[i], pgon1[i+1]])
s1.append([pgon1[-1], pgon1[0]])

s2 = []
for i in range(len(pgon2)-1):
    s2.append([pgon2[i], pgon2[i+1]])
s2.append([pgon2[-1], pgon2[0]])

ps1 = [Segment(i, Point(s1[i][0][0],s1[i][0][1]),
            Point(s1[i][1][0],s1[i][1][1]), 0)
       for i in range(len(s1))]
ps2 = [Segment(i+len(s1), Point(s2[i][0][0],s2[i][0][1]),
            Point(s2[i][1][0],s2[i][1][1]), 1)
       for i in range(len(s2))]
s = ps1+ps2

ints = overlay(s, D)

hl = D.hedges
cycles = []
while len(hl) is not 0:     # get all boundary cycles
    c = []
    e0 = hl.pop()
    e = e0
    c.append(e)
    while True:
        print e,
        e1 = e.nextedge
        if e1 is not e0:
            c.append(e1)
            hl.remove(e1)
```

```
49                e = e1
50            else:
51                break
52        cycles.append(c)
53        print
54
55    # create faces and link them to half edges
56    faces = []
57    for c in cycles:
58        f = Face()
59        f.wedge = c[0]
60        for e in c:
61            e.face = f
62            D.hedges.append(e)
63        if f.area() < 0:
64            f.external = True
65        else:
66            f.external = False
67        faces.append(f)
68    D.faces = faces
```

For our example data in Figure 3.5, the cycling of boundaries for each face results in the following cycles of half-edges for each polygon after the overlay:

```
(9,3)->(10,2)  (10,2)->(1,1)  (1,1)->(3,3)  (3,3)->(4,8)
  (4,8)->(8,8)  (8,8)->(9,3)
(9,3)->(6,6)  (6,6)->(3,3)  (3,3)->(9,3)
(3,3)->(6,6)  (6,6)->(9,3)  (9,3)->(8,8)  (8,8)->(4,8)
  (4,8)->(3,3)
(3,3)->(1,1)  (1,1)->(10,2)  (10,2)->(9,3)  (9,3)->(3,3)
```

3.3 Notes

It has long been recognized in the GIS literature that it is important to keep the information of both maps when we overlay them so that the resulting map will have information from the original data. Frank (1987), for example, identified three types of information that must be stored in the result of a map overlay: non-geometric properties, spatial identifiers such as the key value of each polygon, and geometric descriptions of objects such as their dimension as points, lines, or polygons. The use of the DCEL described above can efficiently achieve all these goals. In this book, we focus only on the geometric aspect. However, given the algorithms described so far, it should not be a stretch to include the other aspects of overlay. More details on polygon overlay are also provided in the excellent book by de Berg et al. (1998).

How efficient are the sweep line algorithm (Bentley and Ottmann, 1979) and the overlay algorithm? The line segment intersection algorithm has a complexity of $O(n \log n + k \log n)$ where n is the number of line segments and k is the number of intersections.

In some special cases, we can expect faster performance. For example, if our aim is simply to check whether there is an intersection in the line segments, Shamos and Hoey (1976) developed an algorithm that can return an answer in $O(n \log n)$ time. When there are two sets of line segments, for example red and blue, and segments in each set do not intersect with themselves, a trapezoid sweep line algorithm was developed by Chan (1994) to compute all the intersections in $O(n \log n + k)$ time. The parallelization of intersection detection has also been noted in the literature (Franklin et al., 1989; Healey et al., 1998).

3.4 Exercises

1. We have mentioned that the Bentley–Ottmann algorithm is theoretically more efficient than the brute-force approach. Now it is time for us to explore this empirically. Write a Python program to randomly generate different numbers of line segments and use both approaches to test how they run.

2. We know that the shapefile has become the most popular file format for storing geospatial data. However, a shapefile does not contain any topological information. In other words, the information in shapefile itself does not tell us how the features in each shapefile are connected. The DCEL data structure introduced in this chapter certainly gives us that. Read the tutorial information about GDAL/OGR in Appendix B and write a Python program that can be used to convert a polygon shapefile into a DCEL format.

3. The `overlay` function developed in this chapter is incomplete. There are two cases for which we have not provided coding. Complete the coding and test the entire algorithm using more complicated input data.

Part II

Spatial Indexing

4

Indexing

Erase the Root – no Tree –
Thee – then – no me –
The Heavens stripped –
Eternity's vast pocket, picked –

Emily Dickinson, *Poem 587*

There are many cases when we need to find a subset of spatial information for various purposes. For example, to interpolate the measure of a variable at a location where we do not have the observation, we need to borrow the observed values from a number of nearby locations. More generally, this can be done when we want to know about the points near any given location. In another example, to quickly pull up information about the polygon we clicked on the map, we need to find the polygon that contains the click point and ignore the many other polygons that geometrically have nothing to do with the point. So, is it a big deal to do this quickly? In other words, does speed matter in this case? By way of analogy, let us try to find the word "indexing" in this book. One may say it is here, on this very page. That is correct. But how do we know this? If we close the book, can we still pinpoint this page so quickly? How about other words? If we know that the word "indexing" is indeed in this chapter, it will be easier because we then know it is not going to be in any of the previous pages (so we can ignore them). In this way, we are indexing the information: we associate the location of the word "indexing" in this case with the number of a chapter, and the chapter numbers are sorted in a specific sequence starting from 1. We can also alphabetically sort all the words (at least the important ones) and associate a page number with each of the words in the sorted list. This is exactly what we have in the Index of this book where we can quickly find the word "indexing" because we can skip words that start with A, B, C, and so on. When we talk about things that happened in the past, we try to link them with the years or our ages, and in this way we treat the events as a time sequence and index the events with time.

The power of indexing lies in a commonly used strategy humans have used for thousands of years: divide and conquer. For example, when we search for the word "indexing" in this book's Index for the first time, we do not know where the I section is located. What we can do is quickly go to the middle of the index list and see if the

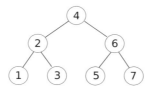

Figure 4.1 A binary tree that stores and sorts the numbers 1, 2, ..., 7

I section is there. If what we find there is K, we would know there is no need to go to the section after that because section I is not going to be there. Now we may go to the end of the first quarter of the list and see if that section is before or after I. Roughly we will do this a few times and thus narrow down our search. Effectively we divide the whole search area into two halves and ignore the half when our target is definitely not there. We keep on dividing the search space into halves until we find what we need.

Indexing can happen in a variety of ways. The phone book is sorted alphabetically, but events are sorted numerically by time. Sometimes when we have a lot of information listed in a table, we may be able to choose how to sort the information. For example, expedia.com allows us to sort flight information by price, departure time, and arrival time, in a descending or ascending order. To accommodate such diverse data sorting requirements in computers, we abstract all kinds of sorting and indexing using a tree structure. We saw how a tree can be used to sort line segments when we discussed line segment intersection in Chapter 3. In this chapter, we introduce some basic features of trees before we discuss how tree structures can be useful for indexing spatial information.

Figure 4.1 shows a tree structure. A tree typically has a root, which most of the time is the main, if not the only, access to the tree. In other words, when we search for information using a tree, we always start at the root. In our case, the root is the node that contains 4. A tree has a set of nodes where information is stored and there is a hierarchy to organize the nodes. Node A is called the parent node of B if A is immediately higher than B in the hierarchy, and B is called the child node of A. The root of a tree of course does not have a parent node. There are nodes in a tree that do not have child nodes, and we call them leaf nodes. In this example, nodes 1, 3, 5, and 7 are leaf nodes. A node and all its child nodes together are called a subtree of the original tree.

The tree in Figure 4.1 is a special tree structure called a binary tree because each node can only have up to two child nodes. In this particular example, we use the tree to represent a number sequence. For each node, we make sure that the node in the left branch holds a smaller number than the number at the node, and the right node holds a higher number. There is a path from the root to each of the nodes and the number of nodes on that path is called the depth or height of that node.

The depth of the entire tree is dependent on the number of nodes on the deepest path of the tree. In Figure 4.1, all the leaf nodes have exactly the same depth. This is a perfectly balanced tree. Perfectly balanced trees are hard to achieve: there must be exactly $2^{H+1} - 1$ nodes to make this happen, where H is the depth (or height) of the tree. However, we can try to create balanced trees where the height difference between the left and right subtrees of each node is not greater than 1.

To use a binary tree to search for a number, we start at the root and if the root is the number to be found then we stop and report the number. Otherwise, we compare the

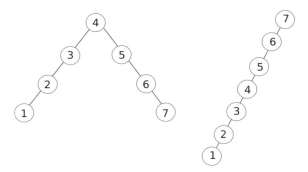

Figure 4.2 Two unbalanced trees for storing 1, 2, ..., 7. The right tree is obviously not balanced. In the left tree, the node 3 has a depth of 3 in the left branch but 0 in the right (so the difference is greater than 1)

number to be found with the root to decide on which branch to continue the search. If the number is smaller than the root, then we go to the left branch and there is no need to look at anything in the right branch. Similarly, if the number is greater we will choose the right branch to continue and ignore everything in the left branch. We continue this process by comparing the number with each node until either we find the number in the tree or the number does not exist in the tree.

Clearly, the search process will take a short amount of time if the tree has a small depth. The tree in Figure 4.1 is an optimal case for the seven numbers because there cannot be a tree with a depth smaller than this. As a matter of fact, there can be many trees that are much worse. Figure 4.2 shows two other example binary trees based on the same data. Neither of these two trees is balanced. Searching for a number in these two trees will in general take more steps to go down the tree and therefore will take more time.

Having developed the idea that data can be indexed as a tree, can we just apply that to spatial data? The answer is yes, but there is something to consider: indexing spatial information is different from other indexing we do in our daily life where things are recognized in one dimension. Objects in space have higher dimensions: we often talk about two dimensions, but we can extend that to a third dimension as in height or elevation. The easiest way to index spatial information is simply to focus on one dimension. For example, we can sort the seven points in Figure 4.3 according to their horizontal axis readings, which gives the sequence of points B, A, E, C, D, F, G. If we want to extract the points that are within a region bounded by a rectangle, we can use the sequence of points to narrow down those with X coordinates in the range of the rectangle. Because the X coordinates are sorted, this can be done quickly using the divide-and-conquer strategy we mentioned before. However, this sequence is only helpful when we want to search for the X coordinates of these points. The Y coordinates are often independent of X and therefore finding points that are in a certain range of X does not exclude the points. For example, the X coordinate of point A is smaller than all other points except B. But this does not tell us whether the Y coordinate of A is smaller or greater than the others. So, we either use the brute-force method to exhaustively search for those points with the desired X coordinates, or we sort the points using their Y coordinates.

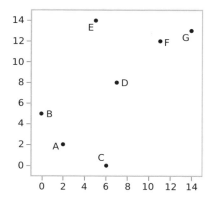

Figure 4.3 A hypothetical data set with seven points

To implement the idea of a binary tree, we first define a data structure to hold information for a node in the tree. Listing 4.1 is a Python class for this purpose. Because Python does not check types, members in this class can actually be of any type, as long as we have a way to compare between different members.

Listing 4.1: A node class for a binary tree.

```
class node():
    def __init__(self, data, left, right):
        self.data = data
        self.left = left
        self.right = right
    def __repr__(self):
        return str(self.data)
```

As already mentioned, everything we do with the binary tree starts from the root. In other words, a tree is actually seen as a node, from which we can access every other node. With this in mind, we can search the information in a tree in a recursive manner using the following Python function (Listing 4.2). Here, the tree is given simply as a node t, and the information to be searched for is d. The Boolean variable is_find_only specifies whether the search is to find the value (True) or to find the immediate parent node of a value that is not present in the tree (False). The latter case is useful when we try to decide where to put a new node – it must be the node that is the immediate parent node of the new node. The search goes to either the left or right branch of the node, depending on the data of the node and the data to be found (lines 4–7). The search function proceeds by continuously calling itself (line 15) until the process reaches the end of the node (line 2), the node that contains the data (line 8), or the node that contains the data does not exist (line 13).

Listing 4.2: Searching the binary tree.

```
def search_bt(t, d, is_find_only=True):
    if t is None:
        return
    if d<t.data:
        next = t.left
    else:
        next = t.right
    if t.data==d:
        if is_find_only:
            return t
        else:
            return
    if not is_find_only and next is None:
        return t
    return search_bt(next, d, is_find_only)
```

We can now write a function that can be used to insert a new node into a tree (Listing 4.3). The code first finds the potential parent node of the new node we are going to insert (line 2) using the search function described above. If the search returns None, indicating the node already exists in the tree, we do not need to do anything (lines 2 and 3). Otherwise, we create a new node (line 5) and assign the new node to either the left (line 7) or right (line 9) of the parent node, depending again on the data.

Listing 4.3: Insert function for a binary tree.

```
def insert(t, d):
    n = search_bt(t, d, False)
    if n is None:
        return
    n0 = node(d, left=None, right=None)
    if d<n.data:
        n.left = n0
    else:
        n.right = n0
```

Now, given a data set, we can create a tree. We assume that each value only appears once in the data as the tree only stores each value once. Listing 4.4 provides a code segment that creates a binary tree for a list of integers. Here we first define a function bt that returns the root of the newly created tree. We also define a function that prints out the data in the tree in ascending order (line 7). At the end of the listing, we shuffle the elements in the list data (line 22) and create a new tree from the randomized data (line 23). The output of the bt_print function should still be the same.

Listing 4.4: Testing the binary tree.

```
def bt(data):
    root = node(data=data[0], left=None, right=None)
    for d in data[1:]:
        insert(root, d)
    return root

def bt_print(t):
    if t.left:
        bt_print(t.left)
    print t
    if t.right:
        bt_print(t.right)

data = range(10)
t = bt(data)
search_bt(t, 0)
bt_print(t)
search_bt(t, 10)
insert(t, 10)
bt_print(t)
import random
random.shuffle(data)
t1 = bt(data)
bt_print(t1)
```

The code presented here is not designed to create a balanced binary tree, which requires more considerations and operations when a node is to be inserted. However, the framework used here will be useful when we discuss the trees for spatial data in this part of the book. In the rest of this part, we introduce a few tree structures that can be used effectively to index spatial data, mainly points and polygons. All these methods utilize the two or more dimensions of spatial information in an integrated manner so that all the dimensions contribute to reducing the search time. It will be noted that, although spatial data are different from one-dimensional data, the principles discussed here are still the same. We will see how the code resembles what we have covered in this chapter.

4.1 Exercises

1. Design a set of computational experiments so that we can compare the search on a binary tree with a linear search.

5

k-D Trees

Nothing is particularly hard if you divide it into small jobs.

Henry Ford

The *k*-D trees are a family of tree data structures specifically designed to handle high-dimensional data. The symbol *k* in the name refers to the dimension of the data. Here, since we are dealing with spatial data in a two-dimensional domain, we focus on 2-D trees. While all the spatial data structures we discuss in this part of the book are two-dimensional, the idea of the *k*-D tree can be used for higher dimensions. There are many kinds of *k*-D trees, and in this book we will focus on two of the basic types: point *k*-D trees and point region *k*-D trees.

5.1 Point *k*-D trees

Let us use the seven points listed in Table 5.1 to illustrate how a point *k*-D tree works. These points are also marked in the left-hand part of Figure 5.1. In this case, we assume the points are entered into the tree in the order $A, B, \ldots, G$ – we will see this order of points has a critical consequence later. Table 5.1 shows the coordinates of these points. To start, we use the first point, A, to divide the space filled by all the points into two parts – the left and right halves – using the X coordinate of the point. In Figure 5.1, we draw a vertical line to indicate the boundary of the two halves. We also create the root of the 2-D tree (since point A is the first point we use) with two branches. Then we examine the second point in the data set, which is B. This time, we use the Y coordinate of B to divide the half plane where B is located into two parts, the upper and lower parts, as indicated by the horizontal line that passes through B. To insert point B into the tree, we will start at the root (where the first point, A, is). Since B has a smaller X coordinate than A, we go to the left branch of the root. The node is empty at this moment and we place point B there. Point B will have two empty branches that can be used for new points that may be inserted later. To insert point C, we again start from the root. Since C has a greater X coordinate than A, we move to the right branch of the root, where the node is empty, and we put point C in the node. In this figure, we draw another horizontal line to divide the half plane of C into the upper and lower parts. We will repeat this process until all the points are inserted in the tree.

Table 5.1 Points used for spatial indexing

Point	X	Y
A	2	2
B	0	5
C	8	0
D	9	8
E	7	14
F	13	12
G	14	13

It is clear from the above description that we alternate the use of X and Y coordinates. The convention in the literature is to use the X coordinate first, and then Y, then X, and so on. The drawing on the left in Figure 5.1 is only for illustration purposes. The tree is searched by comparing the corresponding coordinates at each level of the tree, where the X coordinate is always used first at the root.

Imagine that we wish to find information about point F on a map showing the seven points, and that we click on the point. To find the information we want, we start at the root of the k-D tree. Since the point has a greater X coordinate than the root, we search the right-hand branch of the root, which is point C. This is not our target and we continue to search by comparing, this time, the Y coordinate of the target with that of point C. We go to the right-hand branch since the target has a greater Y coordinate. This leads us to the node containing point D in the tree, and now we use the X coordinate to determine the search direction. We go to the right-hand branch and stop at node F in the tree. If we say that the depth of the root is 0, then we use the X coordinate for all the even depths and Y for odd depths. Throughout the process, we are able to exclude potentially half of the branches at each node, which could make the search efficient.

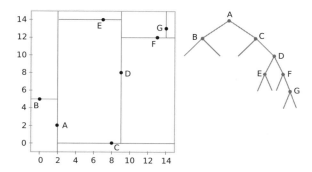

Figure 5.1 A 2-D tree for a set of seven points using the fixed sequence

The code in Listing 5.1 implements some of the ideas discussed above.[1] Here we first define a data structure that holds the necessary information for each node (line 5) by storing three pieces of information: the point for the node, and the left and right nodes.

---

[1] Part of this listing is adopted from http://en.wikipedia.org/wiki/K-d_tree, which is used elsewhere, for example https://code.google.com/p/python-kdtree/. A more comprehensive Python package for k-D trees can be found in SciPy (http://www.scipy.org).

The `kdcompare` function (line 13) returns which branch to follow during a search, given the root node of the tree, the target point, and the depth that determines which coordinate to use.

To insert a new node into a tree, we need a search function that queries the current tree and finds the correct branch where the new point can be placed as a node. The function `query_kdtree` (line 61) performs this task. This is similar to the search function on a binary tree discussed in the previous chapter. This function can be used to search for the point in a tree and to find a node where a new node can be inserted as a child node. The function continuously traverses the tree, guided by the `kdcompare` function. A Boolean variable `is_find_only` is used to indicate whether the goal is to find an empty node that can store the target point (True) or just to find if there exists a point in the tree that has exactly the same location as the target point (False). To find the empty node, the function checks if the child of the current node is empty (line 71). If so, it returns the current node and the branch of the current node that can be used to hold the new point (line 73). Otherwise, if the goal is to use this function to find a point in the tree, it returns the node that contains the point if it exists, or `None` if it does not (line 75). This is a recursive function that stops only when these conditions are met.

With the above functions defined, we now introduce a way of creating the *k*-D tree using the fixed sequence of points. The procedures in function `kdtree` are used to achieve this (line 30). Here we first create the `KDTreeNode` instance that will be used as the root of the tree (line 34). Then, for each input point, we create a new node that includes the point (line 36). We next find the node that has an empty child (line 37) and assign the new node to the correct branch of the parent node (lines 38–41). The tree created in this way is exactly as shown in Figure 5.1. Even a casual look at the tree can tell us that the right branch of the root is much higher than the left branch. This makes the tree unbalanced, which, in turn, increases the time used to search for spatial information because going to the right-hand side of the *k*-D tree does not effectively exclude points to explore.

The unbalanced *k*-D tree is caused by the use of a fixed sequence of points where the order is arbitrarily determined. One way to address this issue is to make a different sequence of points so that every time a point becomes a node in the tree, it sits in the middle of the rest of the points that are to be inserted. The function `kdtree2` (line 45) is set up for this purpose, where each time we use the median point as the node. Here, we first determine the appropriate dimension or axis given the depth (line 49). The coordinates of this axis are then used to sort the points (line 50) and to obtain the median point (line 51). In a recursive fashion (line 55), we use the median point as the new node and we divide the other points into two parts using the median point so that the points with smaller coordinates are used to create the left branch of the current node, and the greater coordinates for the right branch. The process continues until all the points are used (lines 46–47). Using this algorithm on the seven points in our example, we are able to create a perfectly balanced 2-D tree (Figure 5.2).

The query function in Listing 5.1 only does a limited query job by searching for the exact point in (or not in) the tree. In practice, we may wish to query the data in many different ways. For example, we may draw a circle or rectangle on the screen and expect to retrieve all the points that fall within the circle or rectangle. This is what we often refer to as a (circular or orthogonal) range query. Also, we may be interested in looking at the nearest few points to a given point, which may be our current location or any location we can point to on a map.

Listing 5.1: Implementation of the *k*-D tree (kdtree1.py).

```
import sys
sys.path.append('../geom')
from point import *

class kDTreeNode():
    def __init__(self, point, left, right):
        self.point = point
        self.left = left
        self.right = right
    def __repr__(self):
        return str(self.point)

def kdcompare(r, p, depth):
    """
    Returns the branch of searching on a k-d tree
    Input
        r: root
        p: point
        depth : starting depth of search
    Output
        A value of -1 (left branch), or 1 (right)
    """
    k = len(p)
    dim = depth%k
    if p[dim] <= r.point[dim]:
        return -1
    else:
        return 1

def kdtree(points):
    """
    Creates a k-d tree using a predefined order of points
    """
    root = kDTreeNode(point=points[0], left=None, right=None)
    for p in points[1:]:
        node = kDTreeNode(point=p, left=None, right=None)
        p0, lr = query_kdtree(root, p, 0, False)
        if lr<0:
            p0.left = node
        else:
            p0.right = node
    return root

# always use the median point to split the data
def kdtree2(points, depth = 0):
    if len(points)==0:
        return
    k = len(points[0])
    axis = depth % k
    points.sort(key=lambda points: points[axis])
    pivot = len(points)//2
```

```
52          while pivot<len(points)-1 and\
53                  points[pivot][axis]==points[pivot+1][axis]:
54              pivot += 1
55          return kDTreeNode(point=points[pivot],
56                            left=kdtree2(points[:pivot],
57                                         depth+1),
58                            right=kdtree2(points[pivot+1:],
59                                          depth+1))
60
61      def query_kdtree(t, p, depth=0, is_find_only=True):
62          if t is None:
63              return
64          if t.point == p and is_find_only:
65              return t
66          lr = kdcompare(t, p, depth)
67          if lr<0:
68              child = t.left
69          else:
70              child = t.right
71          if child is None:
72              if not is_find_only:
73                  return t, lr
74              else:
75                  return
76          return query_kdtree(child, p, depth+1, is_find_only)
77
78      if __name__ == '__main__':
79          data1 = [ (2,2), (0,5), (8,0), (9,8),
80                    (7,14), (13,12), (14,13) ]
81          points = [Point(d[0], d[1]) for d in data1]
82          p = points[0]
83          t1 = kdtree(points)
84          t2 = kdtree2(points)
85
86          print [ query_kdtree(t1, p) for p in points ]
87          print [ query_kdtree(t2, p) for p in points ]
```

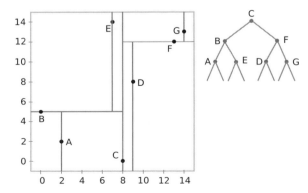

Figure 5.2 A 2-D tree for a set of seven points using the median point at each depth

5.1.1 Orthogonal range query

The orthogonal range query algorithm finds points falling within a rectangle specified by the user. The basic idea here is to find all the possible points that meet the requirements. In doing so, we may need to traverse the entire tree. However, we also want to ignore points that cannot be in the range. In the example shown in Figure 5.3, we want to retrieve precisely point C from the data using the dotted rectangle. It is clear that all the points on the left side of the root (point A) should not be considered because they fall outside the rectangle. But we should continue to check points that are on the right side of the root node. We can extend this observation to all the nodes at any depth. For example, at node G, we can ignore the left subtree of this node because all the points on this subtree must have Y coordinate smaller (lower) than the smallest Y coordinate of the rectangle. However, we must consider the other subtree that is on the right branch of node G, even though for our case the only point on this subtree (I) is not in the rectangle. In general, if the coordinate of the node on the currently searched axis is smaller than the lower bound of the rectangle at the same axis, we can ignore the left subtree of the node but must search the right subtree. If the currently searched axis of the node is greater than that of the upper bound of the rectangle at the same axis, we only need to search further the left subtree of the node, and not the right subtree.

This orthogonal range query algorithm discussed above is implemented in the `range_query_orthogonal` function in Listing 5.2. We use a two-dimensional list (line 27) to represent a search rectangle where the first inner list holds the lower and upper bound in the X coordinate of the rectangle, and the second the Y coordinate (line 27). Line 8 checks if the node point is smaller than the lower bound of the rectangle. We only need to explicitly test if a point is in the rectangle when all the conditions discussed above have been checked (line 16). We use a rectangle to demonstrate the case in the left of Figure 5.4 where two points will be reported:

```
[(2, 2), (9, 8)]
```

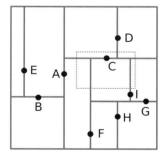

Figure 5.3 Orthogonal range query on a *k*-D tree using a rectangle (dotted)

Listing 5.2: Orthogonal range querying the *k*-D tree (kdtree2a.py).

```
from kdtree1 import *

def range_query_orthogonal(t, rect, found, depth=0):
    if t is None:
        return
    k = len(t.point)
    axis = depth%k
    if t.point[axis] < rect[axis][0]:
        range_query_orthogonal(t.right, rect, found, depth+1)
        return
    if t.point[axis] > rect[axis][1]:
        range_query_orthogonal(t.left, rect, found, depth+1)
        return
    x, y = t.point.x, t.point.y
    if not (rect[0][0]>x or rect[0][1]<x or
            rect[1][0]>y or rect[1][1]<y):
        found.append(t.point)
    range_query_orthogonal(t.left, rect, found, depth+1)
    range_query_orthogonal(t.right, rect, found, depth+1)
    return

def test():
    data1 = [ (2,2), (0,5), (8,0), (9,8), (7,14),
              (13,12), (14,13) ]
    points = [Point(d[0], d[1]) for d in data1]
    t1 = kdtree(points)
    rect = [ [1, 9], [2, 9] ]
    found = []
    range_query_orthogonal(t1, rect, found)
    print found

if __name__ == '__main__':
    test()
```

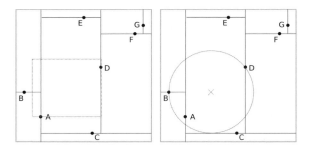

Figure 5.4 Range query of a *k*-D tree of the seven sample points using a rectangle (left) and a circle (right)

5.1.2 Circular range query

The circular range query follows a logic similar to that behind the orthogonal version, but this time we want to find the points that fall within a circle of radius r around a target point p (Listing 5.3). At each node, the algorithm checks if it is necessary to go further down the tree in an effort to avoid exhaustively searching the entire tree. This is done in a recursive manner in the `range_query_circular` function. If the lower bound (of the X or Y coordinate, depending on the depth of the current node in the tree) of the circle is to the right or above the node point (line 7), we need to consider the child node on the right branch (line 8), but we can ignore the child node on the left branch of the node because it is definitely outside the bounds. Similarly, if the upper bound (of the X or Y coordinate) is to the left of the current node, we still need to make sure to consider the child node on the left branch of the node (lines 10 and 11). Otherwise, we know the point at the node is within the bounds (in X or Y) of the circle and we test if it is actually within the circle (line 13) to determine whether we should include it in the result (line 14). Then, we make sure all the child nodes of the node are tested (lines 15 and 16). The process continues until all the points are either tested (lines 5 and 6) or can be ignored (lines 9 and 12). Running the `test` function (line 29) will return three points that are within 5 units of the point (5, 5):

```
[(2, 2), (0, 5), (9, 8)]
```

Listing 5.3: Circular range querying the *k*-D tree (kdtree2b.py).

```
from kdtree1 import *

# range search for points within a radius of r around p
def range_query_circular(t, p, r, found, depth=0):
    if t is None:
        return
    if kdcompare(t, Point(p.x-r, p.y-r), depth)>0:
        range_query_circular(t.right, p, r, found, depth+1)
        return
    if kdcompare(t, Point(p.x+r, p.y+r), depth)<0:
        range_query_circular(t.left, p, r, found, depth+1)
        return
    if p.distance(t.point) <= r:
        found.append(t.point)
    range_query_circular(t.left, p, r, found, depth+1)
    range_query_circular(t.right, p, r, found, depth+1)
    return

def test():
    data1 = [ (2,2), (0,5), (8,0), (9,8), (7,14),
              (13,12), (14,13) ]
    points = [Point(d[0], d[1]) for d in data1]
    p = Point(5,5)
    t1 = kdtree(points)
    found = []
    range_query_circular(t1, p, 5, found)
    print found

if __name__ == '__main__':
    test()
```

5.1.3 Nearest neighbor query

To search for n nearest neighbors of a given point p, we follow a logic similar to that behind the range query: we want to ignore the points that cannot possibly be considered as belonging to those nearest points, and we also do not want to miss any points that should be considered. In other words, we want to find precisely the n nearest points of p. Specifically, at each node that is not excluded from consideration, we check if this node point is nearer to p than the points we have already found. If so, we will update the neighbors by inserting it into the found neighbors. When inserting a new point, we make sure the found neighbors are sorted and we maintain the maximum distance (`maxdist`) between the found neighbors and p. If we have not found n neighbors, the maximum distance will be set to infinity. We then decide which branch of the node is nearer to p and we must make sure the nearer child node is further considered as a possible member of the nearest neighbors. If the distance between the node point and p is smaller than the maximum distance found so far, then there is a need to check the other branch to make sure no points are missed on that side of the space. At any point during the search, if we reach a leaf point, we will check if it should be a member of the nearest neighbors. The process continues until there are no more points to consider.

The algorithm implemented in Listing 5.4 executes the idea discussed above. In order to maintain a list of nearest points found so far, we define a constant of infinity. Because there is no real infinity in a computer (where numbers can be huge but finite), we use a huge built-in number in Python for this purpose (line 3). The function `update_neighbors` maintains a list of points found so far by checking if a new point p0 should be inserted. We organize the neighbors by recording each of the points along with its distance to p in a list called `neighbors`. Here we use a Python built-in list function called `enumerate` to return the index of each element of the list along with the element itself during the enumeration (line 8). If the last element is reached (line 9), the function returns the distance of the last element to p (line 10). Otherwise, if the distance between the new point p0 to p is smaller than that of a member in `neighbors` (line 11), we insert it into the list (line 12). If the total number of neighbors found is smaller than n, we return infinity as the maximum distance (line 14). Otherwise, we return the distance of the nth element in `neighbors` (line 15). When the list is empty, we simply add the new point into the list and return infinity as the maximum distance (lines 16–17).

The core search procedure is done in the `nnquery` function (line 19).[2] The search will stop going further down when the node is empty (line 21) or when it is a leaf (line 23). For a leaf node, we check if the node point should be added into the nearest neighbors (line 24). We go down the tree by tracking the depth, which is in turn used to decide which axis will be used to determine the branch to go (line 26). The nearer and farther subtrees are decided by checking the coordinate of the corresponding axis (lines 28 and 30). We will go down the nearer subtree (line 31) to find if there are more points that are closer to p than the ones that have been found so far. However, this does not mean we will totally ignore the farther subtree. We illustrate this issue in Figure 5.4 where the target point p is at (5, 5). At the root, the right side of the tree is the nearer subtree and the algorithm will go on to explore all the points in this branch. However, the left side of the tree is worth considering. We determine whether the farther side is worth

---
[2]This code is adopted from `https://code.google.com/p/python-kdtree/`.

considering by checking if the distance from the point at the node of the subtree to the target is smaller than the current maximum distance. If it is, then we will also go to the farther subtree. This is executed in line 33.

Listing 5.4: Querying nearest neighbors using the *k*-D tree (kdtree3.py).

```
from kdtree1 import *

INF = float('inf')
maxdist = INF

def update_neighbors(p0, p, neighbors, n):
    d = p0.distance(p)
    for i, x in enumerate(neighbors):
        if i == n:
            return neighbors[n-1][1]
        if d < x[1]:
            neighbors.insert(i, [p0, d])
            if len(neighbors) < n:
                return INF
            return neighbors[n-1][1]
    neighbors.append([p0, d])  # first point
    return INF

def nnquery(t, p, n, found, depth=0):
    global maxdist
    if t is None:
        return
    if t.left == None and t.right == None:
        maxdist = update_neighbors(t.point, p, found, n)
        return
    axis = depth % len(p)
    if p[axis] < t.point[axis]:
        nearer_tree, farther_tree = t.left, t.right
    else:
        nearer_tree, farther_tree = t.right, t.left
    nnquery(nearer_tree, p, n, found, depth+1)
    maxdist = update_neighbors(t.point, p, found, n)
    if (t.point.distance(p)) < maxdist:
        nnquery(farther_tree, p, n, found, depth+1)
    return

def kdtree_nearest_neighbor_query(t, p, n=1):
    nearest_neighbors = []
    nnquery(t, p, n, nearest_neighbors)
    return nearest_neighbors[:n]

if __name__ == '__main__':
    data1 = [ (2,2), (0,5), (8,0), (9,8), (7,14),
```

```
44                  (13,12), (14,13) ]
45       points = [Point(d[0], d[1]) for d in data1]
46       p = Point(5,5)
47       t1 = kdtree(points)
48       n = 3
49       nearests = []
50       nnquery(t1, p, n, nearests)
51       print [x[0] for x in nearests[:n]]
52       print [x[1] for x in nearests[:n]]
53       print sorted([p.distance(x) for x in points])[:n]
```

To use the `nnquery` function, we need to prepare an empty list (line 49). However, the functions we implement here do not maintain a cap on the total number of elements found. Therefore, the actual number of points in the list `nearests` can be more than n. We use the Python list slice operation to return up to n elements (lines 51–53). The last line in the listing conducts an exhaustive search to create a sorted list of all the distances. The results are shown below, where we confirm that the nearest neighbor search returns exactly the same results as the brute-force search.

```
[(2, 2), (9, 8), (0, 5), (8, 0)]
[4.242640687119285, 5.0, 5.0, 5.830951894845301]
[4.242640687119285, 5.0, 5.0, 5.830951894845301]
```

5.2 Point region *k*-D trees

The *k*-D tree we discussed in the previous section is based on the partitioning of space using the coordinates of the points in the data. This poses the challenge of maintaining the tree. Any changes to the points will change the tree structure, regardless of how the tree is created. A tree based on partitioning using the median points, for example, will need to be adjusted once a point is deleted or a new point is inserted. The point region

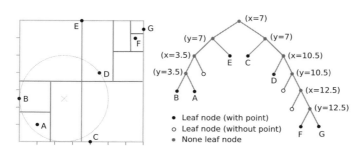

Figure 5.5 A point region *k*-D tree for a set of seven points

(PR) k-D tree addresses this issue by borrowing the idea of the space partitioning approach used in quadtrees, another family of spatial indexing methods (Chapter 6). In a PR k-D tree, we still alternate the use of X and Y coordinates when we build the tree. However, instead of using the coordinates in the point, every time we have to partition the space, we use the center point of the region. Because of this, PR k-D trees have a major difference compared to k-D trees: data points are only held in the leaf nodes of a PR k-D tree (Figure 5.5 right), which is not the case for regular k-D trees (cf. Figure 5.1).

Figure 5.5 shows such a tree using the seven points of our sample data. We again create the tree in a recursive manner by starting at the root, which is determined by the center of the X coordinates. To be able to do this, we will of course need the rectangle that contains the area, and here we can simply use the bounding box of all the points.

Listing 5.5 shows the code that implements the PR k-D tree. The key to storing a PR k-D tree is to explicitly record the information about each region. We take a simple approach here using the lower and upper bounds, or ranges, of the X and Y coordinates of the points (line 9). Lines 81–90 in the code show how the ranges are obtained and used. We also store the center of the area (line 9). The use of the center makes it straightforward to determine on which branch of a node the search should continue, as implemented in the function `prkdcompare` (line 22).

The construction of a PR k-D tree follows the same logic as building the regular k-D tree using the median. However, instead of using the median point that is one of the points in the input data, here we use the center of the region, which is not necessarily a point in the data (and if it is, we still use a separate point as the center, and store the coincident point in a leaf node). The `prkdtree` function (starting line 27) builds a PR k-D tree in a recursive manner. A leaf node is created when there is either one or no point in the region to consider. Otherwise, the algorithm continues to divide the region into two halves using the X or Y coordinate of the center point of the region (lines 41 and 42). The points are sorted by their corresponding coordinate (line 43), and we use the bisect module in Python to find the center value that divides the points into two halves (lines 44 and 45). Then the algorithm needs to compute the bounds and centers of the two subregions (lines 46–54), which will be used to recursively build the left and right branches of the current node (line 55).

The function `query_prkdtree` looks for a certain point in the PR k-D tree, following the same logic as in the k-D tree by continuously going down the tree (lines 71–75) until the node is empty (line 65), is the exact point (line 67), or is an empty leaf node (line 69). We test this algorithm by building a simple PR k-D tree using seven points and test if all the points can be successfully found (line 92).

Lisiting 5.5: Creating and querying a PR k-D tree (prkdtree1.py).

```
1  from bisect import *
2  import sys
3  sys.path.append('../geom')
4  from point import *
5
6  class PRKDTreeNode():
7      def __init__(self, xyrange, center, point, left, right):
8          self.xyrange = xyrange # ranges of xy
```

```
 9              self.center = center    # center coordinate
10              self.point = point      # point
11              self.left = left        # left child subtree
12              self.right = right      # right child subtree
13          def is_leaf(self):
14              return (self.left == None and self.right == None)
15          def __repr__(self):
16              return str(self.center)
17
18   # r-root node of a subtree, p-point, depth-current depth
19   def prkdcompare(r, p, depth):
20       k = len(p)
21       dim = depth%k
22       if p[dim] <= r.center:
23           return dim,-1      # left
24       else:
25           return dim,1       # right
26
27   def prkdtree(points, xylim, center=None, depth=0):
28       if len(points)==1:
29           return PRKDTreeNode(xyrange=xylim,
30                               center=center,
31                               point=points[0],
32                               left=None,
33                               right=None)
34       if len(points)==0:
35           return PRKDTreeNode(xyrange=xylim,
36                               center=center,
37                               point=None,
38                               left=None,
39                               right=None)
40       k = len(points[0])
41       axis = depth % k
42       pmid = (xylim[axis][1]+xylim[axis][0])/2.0
43       points.sort(key=lambda points:points[axis])
44       P = [p[axis] for p in points]
45       pivot = bisect(P, pmid) # includes all <=
46       xmin,xmax=xylim[0][0], xylim[0][1]
47       ymin,ymax=xylim[1][0], xylim[1][1]
48       rangel = [ [xmin, xmax], [ymin, ymax] ]
49       rangel[axis][1] = pmid
50       ranger = [ [xmin, xmax], [ymin, ymax] ]
51       ranger[axis][0] = pmid
52       axis2 = (depth+1) % k
53       pmidl = (rangel[axis2][0]+rangel[axis2][1])/2.0
54       pmidr = (ranger[axis2][0]+ranger[axis2][1])/2.0
55       return PRKDTreeNode(xyrange = xylim,
56                           center = pmid,
57                           point = None,
58                           left=prkdtree(points[:pivot],
```

```
59                                           rangel,pmidl,depth+1),
60                            right=prkdtree(points[pivot:],
61                                           ranger,pmidr,depth+1))
62
63   # query if point p is in t
64   def query_prkdtree(t, p, depth=0):
65       if t is None:
66           return
67       if t.point == p:
68           return p
69       if t.is_leaf():
70           return
71       if prkdcompare(t, p, depth)[1] < 0:
72           next = t.left
73       else:
74           next = t.right
75       p0 = query_prkdtree(next, p, depth+1)
76       if p0 is not None:
77           return p0
78       return
79
80   if __name__ == '__main__':
81       data1 = [ (2,2), (0,5), (8,0), (9,8), (7,14),
82                 (13,12), (14,13) ]
83       points = [Point(d[0], d[1]) for d in data1]
84       px = [p.x for p in points]
85       py = [p.y for p in points]
86       xmin = min(px)
87       xmax = max(px)
88       ymin = min(py)
89       ymax = max(py)
90       xylim = [ [xmin, xmax], [ymin, ymax] ]
91       t = prkdtree(points, xylim)
92       print [query_prkdtree(t, p) for p in points]
```

As with the k-D tree, range and nearest neighbor queries can also be designed on a PR k-D tree. For the range queries, we only focus on the circular range query here (Listing 5.6), and the rectangle orthogonal range query algorithm can be devised similarly. The core function that implements the range query algorithm is `rquery` (line 5). This is a recursive function that stops when the node is empty. At a given node, we get the width and height of the region represented by the node (lines 8 and 9). If the left (or bottom) bound of the region represented by the node is to the right of (or above) the query point on the tree (line 11), then only the right branch of the node needs to be checked for further consideration, while the points on the left branch will not be within the queried range. Similarly, if the right (or upper) bound of the region is to the left of (or below) the query point on the tree (line 15), we only need to check the left branch of the node. In all other cases, there is a need to check if the node point is actually within the search radius (line 18) and check both the left and right branches of the node (lines 21 and 22). The purpose of nesting the function `rquery` inside a wrapping function range_query (line 4) is to allow for returning the list of points found from searching the tree instead of passing a variable to the query function. The algorithm is tested using the same data as in the *k*-D tree and the same query result is returned:

```
[(2, 2), (0, 5), (9, 8)]
```

Listing 5.6: Circular range query using a PR *k*-D tree (prkdtree2.py).

```python
from prkdtree1 import *

# range search for points within a radius of r around p
def range_query(t, p, r):
    def rquery(t, p, r, found, depth=0):
        if t is None:
            return
        w = t.xyrange[0][1] - t.xyrange[0][0]
        h = t.xyrange[1][1] - t.xyrange[1][0]
        if prkdcompare(t, Point(p.x-r-w, p.y-r-h),
                       depth)[1] > 0:
            rquery(t.right, p, r, found, depth+1)
            return
        if prkdcompare(t, Point(p.x+r+w, p.y+r+h),
                       depth)[1] < 0:
            rquery(t.left, p, r, found, depth+1)
            return
        if t.point is not None:
            if p.distance(t.point) <= r:
                found.append(t.point)
        rquery(t.left, p, r, found, depth+1)
        rquery(t.right, p, r, found, depth+1)
        return
    found = []
    if t is not None:
        rquery(t, p, r, found)
    return found

if __name__ == '__main__':
    data1 = [ (2,2), (0,5), (8,0), (9,8), (7,14),
              (13,12), (14,13) ]
    points = [Point(d[0], d[1]) for d in data1]
    px = [p.x for p in points]
    py = [p.y for p in points]
    xmin = min(px)
    xmax = max(px)
    ymin = min(py)
    ymax = max(py)
    xylim = [ [xmin, xmax], [ymin, ymax] ]
    t = prkdtree(points, xylim)
    p = Point(5,5)
    print range_query(t, p, 5)
```

The algorithm for finding nearest neighbors of a given point is implemented in the core function called `prkdtree_nnquery` in Listing 5.7. This algorithm is very similar to that of the *k*-D tree, except for two noticeable differences. First, in line 10, we need to check if the node has a point because not all leaf nodes (or just nodes in general) in a PR *k*-D tree are point nodes. For a *k*-D tree, however, *every* node corresponds to a point in the data. This is not the case for the *k*-D tree, where each node corresponds to the center of a region instead of a point in the data. Second, line 19 is commented out because we only update the list of found points when at leaf nodes. The function `nearest_neighbor_query` wraps the core function and returns the list of points found. The algorithm is then tested using the same data to find three nearest neighbors of the point at (5, 5):

```
[(2, 2), (9, 8), (0, 5)]
[4.242640687119285, 5.0, 5.0]
[4.242640687119285, 5.0, 5.0]
```

Listing 5.7: Nearest neighbor query using a PR *k*-D tree (prkdtree3.py).

```python
from prkdtree1 import *
from kdtree3 import * # use update_neighbors and INF

prmaxdist = INF

def prkdtree_nnquery(t, p, n, found, depth=0):
    global prmaxdist
    if t is None:
        return
    if t.is_leaf() and t.point is not None:
        prmaxdist = update_neighbors(t.point, p, found, n)
        return
    axis,dir = prkdcompare(t, p, depth)
    if dir<0:
        nearer_tree, farther_tree = t.left, t.right
    else:
        nearer_tree, farther_tree = t.right, t.left
    prkdtree_nnquery(nearer_tree, p, n, found, depth+1)
    #prmaxdist = update_neighbors(t.point, p, found, n)
    if (t.center-p[axis]) < prmaxdist:
        prkdtree_nnquery(farther_tree, p, n, found, depth+1)
    return

def nearest_neighbor_query(t, p, n=1):
    nearest_neighbors = []
    prkdtree_nnquery(t, p, n, nearest_neighbors)
    return nearest_neighbors[:n]
```

```
29  if __name__ == '__main__':
30      data1 = [ (2,2), (0,5), (8,0), (9,8), (7,14),
31                (13,12), (14,13) ]
32      points = [Point(d[0], d[1]) for d in data1]
33      px = [p.x for p in points]
34      py = [p.y for p in points]
35      xmin = min(px)
36      xmax = max(px)
37      ymin = min(py)
38      ymax = max(py)
39      xylim = [ [xmin, xmax], [ymin, ymax] ]
40      t = prkdtree(points, xylim)
41      n = 3
42      p = Point(5,5)
43      nearests = nearest_neighbor_query(t, p, n)
44      print [x[0] for x in nearests[:n]]
45      print [x[1] for x in nearests[:n]]
46      print sorted([p.distance(x) for x in points])[:n]
```

5.3 Testing *k*-D trees

How effective is spatial indexing using a *k*-D tree or a PR *k*-D tree? Let us test this using data sets of different sizes. In Listing 5.8, we use a set of random points to test the times used to construct a tree and to search for 100 random points from the data using the tree, compared to a brute-force search.

Listing 5.8: Performance tests of *k*-D trees (kdtree_performance.py).

```
1   from kdtree1 import *
2   from prkdtree1 import *
3
4   import random
5   import time
6   import sys
7   import string
8
9   if __name__ == '__main__':
10      npts = 1000              # default number of points
11      if len(sys.argv)==2:     # user specified number of points
12          if sys.argv[1].isdigit() is True:
13              npts = string.atoi(sys.argv[1])
14      points = []
15      for i in xrange(npts):
16          p = Point(random.random(), random.random())
```

```
            points.append(p)

    time1 = time.time()
    kdt1 = kdtree(points)
    time2 = time.time()
    treet1 = time2-time1

    time1 = time.time()
    kdt2 = kdtree2(points)
    time2 = time.time()
    treet2 = time2-time1

    px = [p.x for p in points]
    py = [p.y for p in points]
    xmin = min(px)
    xmax = max(px)
    ymin = min(py)
    ymax = max(py)
    xylim = [ [xmin, xmax], [ymin, ymax] ]
    time1 = time.time()
    kdt3 = prkdtree(points, xylim)
    time2 = time.time()
    treet3 = time2-time1

    print npts, "|", treet1, treet2, treet3, "|",

    n = 100
    t1 = 0 # time finding 100 points in kdtree
    t2 = 0 # time for balanced kdtree
    t3 = 0 # time for pr kd tree
    t4 = 0 # time for linear search
    pp = random.sample(points, n)
    for p in pp:
        time1 = time.time()
        p1 = query_kdtree(kdt1, p)
        time2 = time.time()
        t1 = t1 + time2-time1

        time1 = time.time()
        p1 = query_kdtree(kdt2, p)
        time2 = time.time()
        t2 = t2 + time2-time1

        time1 = time.time()
        p1 = query_prkdtree(kdt3, p)
        time2 = time.time()
        t3 = t3 + time2-time1

        time1 = time.time()
        for i in range(len(points)):
            if p == points[i]:
```

```
68                     break
69          time2 = time.time()
70          t4 = t4 + time2-time1
71
72      print t1, t2, t3, t4
```

Table 5.2 *k*-D trees performance

	Tree construction			Query			
N	*k*-D	*k*-D*	PR *k*-D	*k*-D	*k*-D*	PR *k*-D	Linear
10000	0.43	0.16	0.56	0.005	0.003	0.009	0.495
20000	0.94	0.47	1.33	0.006	0.004	0.011	1.075
30000	2.02	0.70	2.42	0.008	0.005	0.014	2.025
40000	2.49	0.92	3.21	0.007	0.004	0.011	1.978
50000	2.96	1.05	3.56	0.007	0.004	0.011	2.550
60000	3.19	1.27	4.56	0.008	0.004	0.012	3.300
70000	3.88	1.58	5.52	0.007	0.004	0.012	3.458
80000	4.96	2.37	8.46	0.010	0.006	0.015	5.032
90000	4.89	2.03	6.78	0.007	0.004	0.011	4.452
100000	5.19	2.29	8.10	0.007	0.004	0.011	4.931
200000	11.20	5.40	16.63	0.008	0.005	0.012	10.553
300000	17.95	8.64	25.37	0.010	0.006	0.015	13.693
400000	25.27	12.43	37.33	0.011	0.006	0.015	22.594
500000	33.10	16.38	47.60	0.011	0.007	0.017	27.388
600000	38.14	21.29	59.59	0.011	0.007	0.016	31.133
700000	44.42	24.22	65.11	0.012	0.007	0.019	38.304
800000	52.41	28.89	79.16	0.013	0.008	0.026	44.704
900000	59.60	33.12	98.97	0.013	0.007	0.019	50.814
1000000	71.01	39.90	110.65	0.016	0.009	0.024	57.869

* Balanced

We test for numbers of points ranging from 10,000 to 1 million using the *k*-D and PR *k*-D trees introduced in this chapter, and Table 5.2 shows our results. It should be noted that the absolute time (in seconds) used here is not important because it depends on the kind of computer[3] and programming language used. It is clear that building *k*-D trees takes a significant amount of time, compared with the time spent querying points. The time has an increasing trend as the total number of points to insert into the tree increases (Figure 5.6). When querying the points, the left-hand plot in Figure 5.7 shows the dramatic effect of spatial indexing: all three indexing methods have querying time close to 0, while the linear query approach uses time that linearly increases with the number of points. The right-hand plot of the figure shows the detail plot for the three *k*-D trees.

---
[3]We used a MacBook Air with a 1.3 GHz Intel Core i5 chip and 4 GB memory.

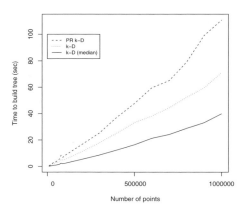

Figure 5.6 Time taken to build *k*-D trees

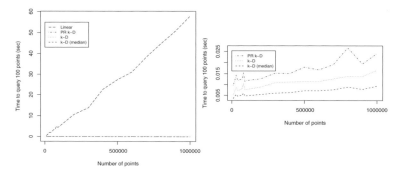

Figure 5.7 Querying time

Comparing the querying time with the time taken to build trees, however, one may ask what is the point of using *k*-D trees if the time required to construct such a tree is even longer than searching using the linear approach? This is especially the case when the total number of points is large. We should make it clear that spatial indexing is not designed for casual uses of spatial data. For example, it is probably not worth the time if all we need is to find a point in the data and will never come back to the same data again. However, there are many real-world applications where the same data are used over and over again, by probably millions of people. Online mapping is a perfect example of this – imagine how many people use Google Maps every day, trying, for example, to find the nearest restaurant. Some research, especially that concerned with the use of individual census data (Young et al., 2009), also requires intensive querying of the data set to obtain point locations. Also, let us not forget there are other query tasks than the simple one tested here. Range querying, for example, will be significantly slower for brute-force search because computing distance is a non-trivial job.

If we focus on the three k-D trees in the above test, it is obvious that the balanced k-D tree using median points outperforms the other two trees. There are no surprises here: the balanced tree is better. But why does the balanced tree also require less time to build? Examining the code shows that, at every step, the balanced k-D uses the default sort algorithm of Python to put the points into order, while the other two trees need to use the existing tree created so far to find the right node; but the trees are not balanced and therefore are not as efficient as the quick sort method.

Building a k-D tree relies on some kind of sorting technique. In our implementations, we either use the built-in Python method to sort the points based on the corresponding axis at each time, or use the order of points that is already determined before we build the tree. Sorting can be done efficiently, but adding the points together is the part that takes much of the running time when we insert points into a tree. The overall time complexity of building a k-D tree is $O(n \log n)$. Searching a k-D tree, however, only takes $O(\log n)$ time in average. However, we can imagine a worst case where points are always aligned on one branch of the node and in this case the search time is $O(n)$, the same as for the linear search.

Theoretically, how much space (memory) do we need to store a k-D tree? This is dependent on the number of nodes we will have to use to create the entire tree. As shown in Figures 5.1 and 5.2, we will need exactly the same number of nodes as the number of points. So the storage is $O(n)$, where n is the number of points. For a PR k-D tree, however, more nodes are needed because we allow for non-leaf nodes and leaf nodes with no data. In the best scenario where all the leaf nodes have data and are at exactly the same depth, we have a perfectly balanced tree with a height of $\ln n$. In this case, the total number of nodes is $n \ln n$.

5.4 Notes

Bentley (1975) first came up with the idea of the k-D tree as a data structure to retrieve information using a binary tree. Since then there have been many new developments in the literature. Hanan Samet has written three excellent books on spatial indexing (Samet, 1990a,b, 2006). These books include not only the classic k-D trees as introduced in this chapter, but also many variants. There are many interesting topics in k-D trees that are not covered in this chapter (see the exercises below). One important idea is the use of buckets. In the k-D trees discussed so far, each node (or region, as in the PR k-D tree) contains at most one point. This is convenient when we are querying the tree since at each node we only need to compare with at most one point. However, associating only one point with a node also makes the size (depth) of the tree unnecessarily large. To address this issue, we can allow multiple points to be associated with a node. This can significantly reduce the tree size (and therefore the time to build the tree). If we keep the bucket size relatively small, searching for points in a bucket will not increase the search time too much. Also, the use of buckets reduces the tree size, which decreases the overall query time.

5.5 Exercises

1. Modify the code of the three query algorithms for the *k*-D tree so that it will report all the points tested during the search process. Do the algorithms ignore the nodes and points as they should logically?

2. Test the performance of the range and nearest neighbor query methods for *k*-D trees. Listing 5.8 includes the necessary code that can be reused to create random point data sets for testing.

3. Having covered our first spatial indexing technique, it is time to see if indexing really increases the performance of a real application. Use a *k*-D tree on one of the interpolation methods (see Chapter 8) and see if there is any significant change in performance.

4. We have discussed the orthogonal range query for the *k*-D tree. The PR *k*-D tree has the same tree structure, but the search process may include some nuances to develop. Can you implement an orthogonal range query algorithm for the PR *k*-D tree?

5. Prove that a balanced binary tree with n leaf nodes has a total of $n \ln n$ nodes on the entire tree.

6. What happens if distance (or spatial information in general) is not the only criterion needed for nearest neighbor query? For example, if each point is associated with a total population and we require the nearest neighbors to have more than 100 people. Modify the code to return the correct results.

7. Modify the code in `kdtree1.py` so that we can allow each node to store up to two points. How will this impact the overall performance of the *k*-D tree on data sets of different sizes?

8. In Chapter 1 we discussed an algorithm to find the shortest distance between two points from a list of points. The algorithm has a running time of $O(n^2)$, where n is the number of points in the list. Can we build a *k*-D tree to significantly improve the efficiency in solving the problem? Write the Python code and also discuss the theoretical running time of your code. Can you prove your answer empirically?

6

Quadtrees

A quadtree uses both X and Y coordinates to index geospatial data, which is a different strategy from the k-D tree where only one dimension is used at a time. In order to use both coordinates, a node in a quadtree partitions the area into four parts, which, depending on different design ideas and rules, might be further partitioned into four smaller parts. The idea of quadtrees can be applied to not only points, but also higher-dimensional geometric objects such as lines and polygons, and different data models such as the raster model. In the GIS literature, the use of quadtrees in raster data has probably received the most attention, arguably due to the fact that raster is a useful data model for a wide range of applications in digital elevation models and remote sensing images. In this chapter we start with region quadtrees, the type of quadtree specifically designed for raster data. Then we discuss quadtrees for point data, called point quadtrees. In the next chapter, we discuss the use of quadtrees for line and polygon data.

6.1 Region quadtrees

For a raster data set, if the entire region is homogeneous, meaning all the cells take the same value, then it is unnecessary to store the value of every cell in a redundant way. Instead, we could simply use one value to represent the whole area. This is precisely the idea behind the concept of the region quadtree: we continuously partition the area into four equal parts until each of the parts is homogeneous. Here, each part is called a quadrant as it is a quarter of the larger region from which it is generated. Throughout the process of partitioning the region, we create a tree at the same time. The root of the tree represents the entire region before it is partitioned. If the entire region is homogeneous, then the root is also a leaf node and is assigned the value of the cells. Otherwise, the root is a non-leaf node and contains four branches leading to four new nodes. A node becomes a leaf node if it represents a homogeneous region. A non-leaf node will be associated with four more child nodes until leaf nodes can be achieved.

    Let us use the hypothetical raster data on a 16 × 16 grid in Figure 6.1A to illustrate the idea of the region quadtree. Without loss of generality, we only have two kinds of values, 0 (white) and 1 (grey), in this example grid. In a real application, there may be more values. Here the entire region is not homogeneous and therefore requires further partitioning.

Our first step is to create a non-leaf node as the root of the tree (i.e., the node at the top of the tree in Figure 6.1), which has four child nodes. Each of the child nodes points to one of the quadrants in Figure 6.1B. To make sure we have a consistent way of recording information, we always order the quadrants clockwise from

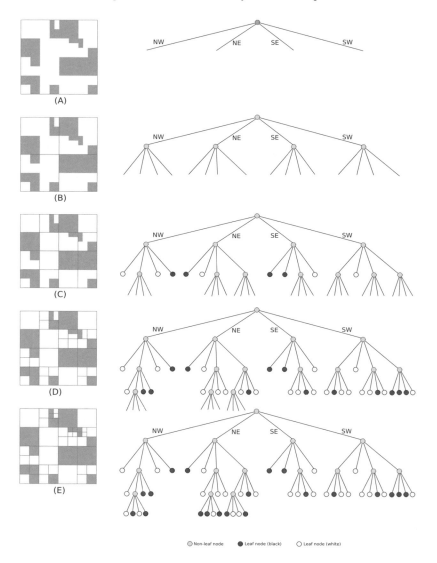

Figure 6.1 Partitioning the region of a raster data set on a 16 × 16 grid into quadrants. The tree to the right of each figure indicates the construction of a region quadtree at the specific step of partitioning. The final tree is presented at the bottom associated with (E)

the northwest (upper right), forming a sequence NW, NE, SE, SW as marked in Figure 6.1. We will use this order for all quadtrees in this book. Since none of these quadrants is homogeneous, each of them is represented as a non-leaf node, which means each will have four more child nodes. We focus on the northwest quadrant for the purposes of discussion. By further partitioning this quadrant into four smaller parts, we can observe that the NW, SE, and SW smaller quadrants are now homogeneous. Therefore these three smaller quadrants are represented as leaf nodes in the tree. The NE smaller quadrant, however, needs to be further partitioned to become homogeneous. This process continues until all the partitions are homogeneous and the final tree can then be created.

The code in Listing 6.1 is the Python program that implements the idea of the region quadtree as discussed so far. Before we look at the code, let us first deal with how to store raster data. Again, we try to use a straightforward and simple approach to representing the data. Here, we use a simple two-dimensional NumPy array to store the data, as shown in the two examples in the code (lines 70 and 76), where the second example is the same as the data in Figure 6.1. We could use Python lists to achieve the same result. However, NumPy provides some very handy tools to slice the array so we can quickly get a "block" from a two-dimensional array (discussed later). This is a great tool because it allows us to focus on the idea instead of programming tricks to make things work.

Given a two-dimensional array, the function `homogeneity` (line 14) is used to test if all the members in the array have the same value. This will be used to determine if the partitioning process has reached a homogeneous quadrant and a leaf node can be created. The mechanism of testing homogeneity is straightforward: we go through all the members of the array and return `False` whenever we find a member that has a different value from the first member.

We define a special data structure called `Quad` (line 3) to store information on each quadrant at each step of the partitioning process. A quadrant is also a node in the tree. Specifically, three pieces of critical information are recorded in a `Quad` instance: the value of the quadrant if it is homogeneous, the dimension or size of the quadrant, and the four child nodes. Here, we do not explicitly mark whether a node is leaf or non-leaf because a leaf node can be easily detected by checking if all the child nodes are empty (`None` in Python). A non-leaf node will also have a `None` value because it is not homogeneous (so we do not know which value to assign).

With the above preparation, we are ready to create a region quadtree with a given data set. This is done using the function `quadtree` (line 22). Here we first figure out the size of the data passed to this function using the `shape` method associated with the `numpy` array. This function returns the number of elements in the specified dimension, and we use it to get the number of rows in the data. The variable `dim` holds the current size and `dim2` holds the size of the child nodes, which is exactly a half of the current size by the design of quadtree. As hinted in Figure 6.1 and the above discussion, the data used to create a quadtree must have the same number of rows and columns. Furthermore, the number of rows (or equivalently columns) must equal 2^n, where n is a constant. If the current data are homogeneous (line 25), the algorithm stops by returning a leaf node as an instance of `Quad` with all child nodes being `None`. Otherwise, it returns the current node without a value and calls itself to create four child nodes with the data sliced from the current data. For example, the

NW quadrant gets the data in rows from 0 to `dim2` and columns from 0 to `dim2`, excluding row or column `dim2` itself in the result of slicing (line 30). The recursive calls will end when all smaller partitions are homogeneous.

Listing 6.1: Region quadtree (quadtree1.py).

```
1   import numpy as np
2
3   class Quad():
4       def __init__(self, value, dim, nw, ne, se, sw):
5           self.value = value
6           self.dim = dim
7           self.nw = nw
8           self.ne = ne
9           self.se = se
10          self.sw = sw
11      def __repr__(self):
12          return str(self.value)
13
14  def homogeneity(d):
15      v = d[0,0]
16      for i in d:
17          for j in i:
18              if j <> v:
19                  return False
20      return True
21
22  def quadtree(data):
23      dim = data.shape[0]
24      dim2 = dim/2
25      if homogeneity(data) is True:
26          return Quad(value=data[0,0], dim=dim,
27                  nw=None, ne=None, se=None, sw=None)
28      return Quad(value=None,
29                  dim = dim,
30                  nw = quadtree(data[0:dim2, 0:dim2]),
31                  ne = quadtree(data[0:dim2, dim2:,]),
32                  se = quadtree(data[dim2:,dim2:,]),
33                  sw = quadtree(data[dim2:,0:dim2]))
34
35  ##############################################################
36  # Note: x, y start at the upper-left corner
37  #
38  # Child node quadrants are ordered as
39  # [q.nw, q.ne, q.sw, q.se]. The side of cell (x, y) in each
40  # region is determined by dx and dy. Then the quadrant
```

```
# (x, y) is located in is calculated as dx+dy*2.
#
#           dx
#
#      ----+-----+------
#          |  0  |  1
#      ----+-----+------
# dy 0 |  0  |  1
#      ----+-----+------
#      1 |  3  |  2
#      ----+-----+------
#
###############################################################
def query_quadtree(q, x, y):
    if q.value is not None: # this is a leaf node
        return q.value      # return the value
    dim = q.dim
    dim2 = dim/2
    dx,dy = 0, 0
    if x>=dim2:
        dx = 1
        x = x-dim2
    if y>=dim2:
        dy = 1
        y = y-dim2
    qnum = dx+dy*2
    return query_quadtree([q.nw,q.ne,q.sw,q.se][qnum], x, y)

def test():
    data0 = np.array(
        [[0, 1, 2, 2],
         [2, 2, 2, 2],
         [0, 0, 2, 0],
         [0, 0, 1, 0]])

    data1 = np.array(
        [[0,0,0,0,0,0,1,0,1,1,1,1,0,0,0,0],
         [0,0,0,0,0,0,1,0,1,1,1,1,0,0,0,0],
         [0,0,0,0,0,0,1,1,1,1,1,0,0,0,0,0],
         [0,0,0,0,0,0,1,1,1,1,1,0,0,0,0,0],
         [1,1,1,1,0,0,0,0,0,0,1,1,1,0,0,0],
         [1,1,1,1,0,0,0,0,0,0,0,1,0,0,0,0],
         [1,1,1,1,0,0,0,0,0,0,0,0,0,1,1],
         [1,1,1,1,0,0,0,0,0,0,0,0,0,1,1],
         [0,0,1,1,0,0,0,0,1,1,1,1,1,1,1,1],
         [0,0,1,1,0,0,0,0,1,1,1,1,1,1,1,1],
```

```
87                [0,0,0,0,0,0,0,0,1,1,1,1,1,1,1,1],
88                [0,0,0,0,0,0,0,0,1,1,1,1,1,1,1,1],
89                [1,1,1,1,0,0,0,0,0,0,0,0,0,0,0,0],
90                [1,1,1,1,0,0,0,0,0,0,0,0,0,0,0,0],
91                [0,0,1,1,0,0,1,1,0,0,0,0,0,0,1,1],
92                [0,0,1,1,0,0,1,1,0,0,0,0,0,0,1,1]])
93
94       q = quadtree(data1)
95       for y in range(q.dim):
96           for x in range(q.dim):
97               print query_quadtree(q, x, y),
98           print
99       return q
100
101  if __name__ == '__main__':
102      test()
```

Given a region quadtree, we can query the value at column x and row y using the function `query_quadtree` (line 54). It is important to note that here we start the row and column numbering at the upper-left corner of the grid. The key is to convert the location specified as a pair of x and y values into the correct quadrant at each step. A straightforward conversion will be needed if a different origin is used. In order to do so, we query the quadtree by adjusting the row and column number at each level of the tree related to the level. We use `dx` and `dy` to indicate which half the querying location is in with respect to the center of the region: `dx` is 0 (1) if it is in the left (right) half horizontally, and `dy` is 0 (1) if it is in the upper (lower) half vertically. The comment code in lines 35–53 provides details about how to determine the `dx` and `dy` values. If x is in the right half of the region, we will need to modify the x value for querying the next level of the tree (line 62), and the same must be done for the y value. Line 66 computes the quadrant number (qnum) of the querying location, which ranges from 0 to 3. The order of child node quadrants is set in the list in line 67, where we use the child node indicated by qnum to recursively search the quadtree.

The function `test` in Listing 6.1 uses the data shown in Figure 6.1 to construct a quadtree. Then we simply make repeated use of the function `query_quadtree` to search for the value at each cell. The output should return exactly the same values as in the example:

```
0 0 0 0 0 0 1 0 1 1 1 1 0 0 0 0
0 0 0 0 0 0 1 0 1 1 1 1 0 0 0 0
0 0 0 0 0 0 1 1 1 1 1 1 0 0 0 0
0 0 0 0 0 0 1 1 1 1 1 1 0 0 0 0
1 1 1 1 0 0 0 0 0 0 1 1 1 0 0 0
1 1 1 1 0 0 0 0 0 0 0 0 1 0 0 0
1 1 1 1 0 0 0 0 0 0 0 0 0 0 1 1
1 1 1 1 0 0 0 0 0 0 0 0 0 0 1 1
```

```
0 0 1 1 0 0 0 0 1 1 1 1 1 1 1 1
0 0 1 1 0 0 0 0 1 1 1 1 1 1 1 1
0 0 0 0 0 0 0 0 1 1 1 1 1 1 1 1
0 0 0 0 0 0 0 0 1 1 1 1 1 1 1 1
1 1 1 1 0 0 0 0 0 0 0 0 0 0 0 0
1 1 1 1 0 0 0 0 0 0 0 0 0 0 0 0
0 0 1 1 0 0 1 1 0 0 0 0 0 0 1 1
0 0 1 1 0 0 1 1 0 0 0 0 0 0 1 1
```

6.2 Point quadtrees

The principles of the quadtree can be extended to indexing points: we continue to partition the space into quadrants until each quadrant contains no more than one point. Let us take Figure 6.2 as an example where the splitting of space is only carried out at each data point. This is what we call a point quadtree; here we use the same seven points as in the previous section. We insert the points $A, B, \ldots, G$ into the tree, with point A being the root. The vertical and horizontal lines passing through point A split the space into four quadrants. Since point B is the only one in the northwest quadrant, the node in the NW branch of the root becomes a leaf node containing point B. A node is a leaf node if it has four empty branches. Whenever a new point is to be inserted, we find the correct branch where the node below is empty. In our case, when point C is to be inserted into the tree, it should go to the SE branch of the root, where the node is empty. In the same way we insert point D into the NE branch of the root. Point E should also be inserted into the NE branch of the root. However, because the node under that branch is already occupied, we compare it to point D and insert it into the node under the NW branch of node D. We continue this process until all the data points are inserted into the tree. Apparently, this process does not guarantee a balanced tree because we use an existing sequence of the points.

We now define a data structure that can be used to store a node of a point quadtree. In Listing 6.2, the class called PQuadTreeNode (line 5) is designed for this purpose: we keep the point along with the four branches of each node. The class also comes with a member function is_leaf to check if a node is a leaf node, which occurs if all the four branches are empty, or None as coded in Python. The function insert_pqtree inserts a point p

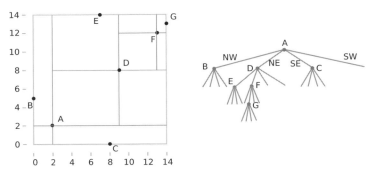

Figure 6.2 Point quadtree

into a point quadtree q (line 37). Here, we first find the correct node that will become the parent of the new node and then decide which branch should be used to hold the new node. Searching for the correct node is achieved using the function called `search_pqtree` by setting the parameter `is_find_only` to False. The search algorithm as implemented in the function `search_pqtree` (line 18) goes down the tree (q) using the relationship between the target point (p) and each of the nodes on the search path. The `point-quadtree` function (line 49) creates a point quadtree by first creating the root and then sequentially inserting each of the remaining points into a tree. At the end of the listing, the same data as shown in Figure 6.2 is used to test the algorithms implemented here.

Listing 6.2: Point quadtree (pointquadtree1.py).

```
1  import sys
2  sys.path.append('../geom')
3  from point import *
4
5  class PQuadTreeNode():
6      def __init__(self,point,nw=None,ne=None,se=None,sw=None):
7          self.point = point
8          self.nw = nw
9          self.ne = ne
10         self.se = se
11         self.sw = sw
12     def __repr__(self):
13         return str(self.point)
14     def is_leaf(self):
15         return self.nw==None and self.ne==None and \
16             self.se==None and self.sw==None
17
18 def search_pqtree(q, p, is_find_only=True):
19     if q is None:
20         return
21     if q.point==p and is_find_only:
22         return q
23     dx,dy=0,0
24     if p.x>q.point.x:
25         dx=1
26     if p.y>q.point.y:
27         dy=1
28     qnum = dx+dy*2
29     child = [q.sw, q.se, q.nw, q.ne][qnum]
30     if child is None:
31         if not is_find_only:
32             return q
33         else: # q is not the point and no more to search for
```

```python
                return
        return search_pqtree(child, p, is_find_only)

def insert_pqtree(q, p):
    n = search_pqtree(q, p, False)
    node = PQuadTreeNode(point=p)
    if p.x<n.point.x and p.y<n.point.y:
        n.sw = node
    elif p.x<n.point.x and p.y>=n.point.y:
        n.nw = node
    elif p.x>=n.point.x and p.y<n.point.y:
        n.se = node
    else:
        n.ne = node

def pointquadtree(data):
    root = PQuadTreeNode(point = data[0])
    for p in data[1:]:
        insert_pqtree(root, p)
    return root

if __name__ == '__main__':
    data1 = [ (2,2), (0,5), (8,0), (9,8), (7,14),
              (13,12), (14,13) ]
    points = [Point(d[0], d[1]) for d in data1]
    q = pointquadtree(points)
    print [search_pqtree(q, p) for p in points]
```

The circular range search algorithm works very similarly to with the *k*-D tree: we are trying to ignore half of the tree when that half is outside the nearest boundary of the bounding box of the search circle. The difference, however, is that we have four branches from each point quadtree node and we must combine the branches on one side of the node point together to form the correct half of the quadrant. Then we determine whether it is necessary to further search each half. When the left bound of the circle is greater than the *X* coordinate of the point at a node, for example, there is a need to further consider all the points on the east (right) side of the node, which would include the NE and SE branches of the node, while the points on the west side (NW and SW) of the node cannot possibly fall within the circle and therefore can be ignored. We check the similar situations for the west, south, and north sides of the bounding box of the circle. For all other nodes, they could fall into the circle and therefore all the four quadrants must be searched for further candidate points. This algorithm is implemented in the Python code shown in Listing 6.3, in which line 10 refers to the situation where we only need to search for all the points on the east side of the node. At the end of the listing we test the same data with a radius of 5 around the point (5, 5), and it returns three points, exactly as in the case of the *k*-D tree.

Listing 6.3: Circular range search on point quadtrees (pointquadtree2.py).

```python
import sys
sys.path.append('../geom')
from point import *
from pointquadtree1 import *

def range_query(t, p, r):
    def rquery(t, p, r, found):
        if t is None:
            return
        if p.x-r > t.point.x :
            rquery(t.ne, p, r, found) # right points only
            rquery(t.se, p, r, found)
            return
        if p.y-r > t.point.y:
            rquery(t.ne, p, r, found) # above points only
            rquery(t.nw, p, r, found)
            return
        if p.x+r < t.point.x:
            rquery(t.nw, p, r, found) # left points only
            rquery(t.sw, p, r, found)
            return
        if p.y+r < t.point.y:
            rquery(t.se, p, r, found) # below points only
            rquery(t.sw, p, r, found)
            return
        if p.distance(t.point) <= r:
            found.append(t.point)
        rquery(t.nw, p, r, found)
        rquery(t.ne, p, r, found)
        rquery(t.se, p, r, found)
        rquery(t.sw, p, r, found)
        return
    found = []
    if t is not None:
        rquery(t, p, r, found)
    return found

if __name__ == '__main__':
    data1 = [ (2,2), (0,5), (8,0), (9,8), (7,14),
              (13,12), (14,13) ]
    points = [Point(d[0], d[1]) for d in data1]
    q = pointquadtree(points)
    p = Point(5, 5)
    print range_query(q, p, 5)
```

Finally, we apply the nearest neighbor query algorithm to a point quadtree (Listing 6.4). The algorithm is implemented in the core function `pq_nnquery`, where the general idea is to make sure the query goes down the tree as deeply as possible, as long as there are still points in the tree that must be checked. Here we use the same function `update_neighbors` from the *k*-D tree to maintain the points found so far (line 2). Again, when we find a point that is near our target, we store the point and its distance to the target point in a list. When the search process reaches a leaf node, we make sure it is considered by comparing it with the nearest points found so far (line 24). For all non-leaf nodes, we use the function `pqcompare` to find the quadrant that is nearest to the target point (line 26) and then go on to the corresponding branch to search for more points (line 28) until all the leaves are reached. Then, if the point on the non-leaf node is closer to the target than the farthest point found so far, we want to make sure that the other quadrants are searched too (code from line 30). Practically, we use a wrapping function `nearest_neighbor_query` (line 36) that conveniently returns a list of the points found, along with their distances to the target point. At the bottom of the code listing, we test the algorithm using the same data and get exactly the same result for a search for the three nearest points around (5, 5):

```
[[(2, 2), 4.242640687119285], [(9, 8), 5.0], [(0, 5), 5.0]]
[4.242640687119285, 5.0, 5.0]
```

Listing 6.4: Nearest neighbor query on point quadtrees (pointquadtree3.py).

```
1  from pointquadtree1 import *
2  from kdtree3 import update_neighbors
3  
4  INF = float('inf')
5  pqmaxdist = INF
6  
7  # returns the quad of t where p is located
8  # 0-NW, 1-NE, 2-SE, 3-SW
9  def pqcompare(t, p):
10     if p.x<t.point.x and p.y<t.point.y:
11         return 3 # sw
12     elif p.x<t.point.x and p.y>=t.point.y:
13         return 0
14     elif p.x>=t.point.x and p.y<t.point.y:
15         return 2
16     else:
17         return 1
18  
19  def pq_nnquery(t, p, n, found):
20     global pqmaxdist
21     if t is None:
22         return
23     if t.is_leaf():
24         pqmaxdist = update_neighbors(t.point, p, found, n)
```

```
25            return
26        quad_index = pqcompare(t, p)
27        quads = [t.nw, t.ne, t.se, t.sw]
28        pq_nnquery(quads[quad_index], p, n, found)
29        pqmaxdist = update_neighbors(t.point, p, found, n)
30        if (t.point.distance(p)) < pqmaxdist:
31            for i in range(4):
32                if i <> quad_index:
33                    pq_nnquery(quads[i], p, n, found)
34        return
35
36   def nearest_neighbor_query(t, p, n=1):
37       nearest_neighbors = []
38       pq_nnquery(t, p, n, nearest_neighbors)
39       return nearest_neighbors[:n]
40
41   if __name__ == '__main__':
42       data1 = [ (2,2), (0,5), (8,0), (9,8), (7,14),
43                 (13,12), (14,13) ]
44       points = [Point(d[0], d[1]) for d in data1]
45       q = pointquadtree(points)
46       p = Point(5,5)
47       n = 3
48       print nearest_neighbor_query(q, p, n)
49       print sorted([p.distance(x) for x in points])[:n]
```

6.3 Notes

Finkel and Bentley (1974) first developed the idea of the quadtree as a data structure to retrieve information. Observe that the name of Bentley has appeared three times so far in this book: the Bentley–Ottmann algorithm for testing line segments intersection, the k-D tree, and in this chapter the quadtrees. As with the k-D trees, a lot of useful details about quadtrees can also be found in Samet (1990a,b, 2006).

The cost of building a point quadtree is $O(n\log_4 n)$ when points are randomly sorted before they are inserted into the tree as we discussed above. A simple search on a balanced point quadtree has a time complexity of $O(n\log_4 n)$, while the worst case would be $O(n)$ when the tree has only one node at each of the levels. It has been shown that the cost of range search would be $O(2n^{1/2})$ for the worst case (Bentley et al., 1977; Lee and Wong, 1977). The study by Lee and Wong (1977) also shows that search for s points from the data would cost $O(sn^{1/2})$ in the worst case.

How do we create a balanced point quadtree? Compared with the k-D tree, a balanced point quadtree is more difficult to construct. Recall that, in the k-D tree, we can alternately use the X and Y coordinates to split the data into two halves using the median point. For a point quadtree, however, it is hard to do the same using the X and Y coordinates together.

One idea is to ensure that at least the tree has approximately the same number of points on the left and right, or upper and lower sides of the node, but not all four sides. To achieve this, we first find the median point according to the X coordinates and use that point to partition the space. At the next level of space partitioning, we sort based on Y coordinates. We alternate these two dimensions until the full tree is constructed.

Another approach to addressing the unbalanced nature of point quadtrees is to split the space at the center instead of at the data point. This is called the point region (PR) quadtree, and we have seen how this idea can be applied to the k-D tree in the previous chapter. If the points are randomly or uniformly distributed in space, this approach becomes very effective since it is likely that the tree will have balanced branches. However, for data points that are highly clustered, this approach may return a highly unbalanced quadtree. We will explore this type of quadtree in the next chapter.

6.4 Exercises

1. We took a parsimony approach to introducing the region quadtree by only focusing on the tree construction and simple querying. However, it should not be difficult to develop a range query on a region quadtree. Review the range algorithm for k-D trees discussed in the previous chapter and design your own range query for a region quadtree. The ideas used in the point quadtree for range query may also be useful in this exercise.

2. For a non-leaf node in a region quadtree, we use None for the value of that node. However, it may be very useful if we know the values in the region represented by that node. Write a Python program to report the values for each node in a region quadtree. This of course requires an exhaustive search of all the branches of that node.

3. In the previous chapter we compared different k-D tree indexing methods using data sets of different sizes. Write a Python program to evaluate point quadtrees using the random data sets from the previous chapter and examine how point quadtrees compare with the other point indexing methods.

4. Write Python code to support the optimization of point quadtrees by alternately using the median points of X and Y coordinates. It would also be interesting to compare the balanced tree with an unbalanced tree.

5. Modify the code in `pointquadtree1.py` so it can print the depth of a quadtree.

6. Design a procedure to conduct rectangle range queries and implement it in Python.

7

Indexing Lines and Polygons

The previous two chapters focused on indexing points (and region quadtrees are designed for raster data, which can of course be treated as regularly placed points). We should have developed the confidence that spatial indexing significantly expedites spatial query, albeit at the initial cost of creating the trees, plus a certain amount of storage to maintain the trees. But overall, the benefit far exceeds the cost for repeated use. In this chapter, we introduce two more indexing approaches, designed for lines and polygons.

7.1 Polygonal map quadtrees

We have seen how quadtrees can be used to index points by partitioning the space as long as there are still points left to be inserted into the tree. Here, we use a different kind of quadtree, called the point region (PR) quadtree, where the space is split at the center of each region instead of at a point. We have seen how this works, in a different context, in PR k-D trees (see Section 5.2), and Figure 7.1 shows such an example. We first partition the space into four quadrants using the center of the entire area. Then we continue to partition each quadrant into smaller quadrants if certain criteria are met. The first criterion is that we continue to split the quadrant as long as there is more than one point. Then the regions are further split when certain line-related conditions are met. Figure 7.1 illustrates a situation where such partitioning occurs if a region contains more than one line segment, unless the line segments share the same vertex that is also in the same region. For example, the lower-right region is not split because it contains two lines that share the same node in the same region. The same applies to the region immediately above, where four lines share the same node. This kind of quadtree is called a polygonal map (PM) quadtree because it can be used to index lines and polygons. Depending on how partitioning is done based on the lines, there are three different versions of PM quadtrees. Figure 7.1 shows a PM1 quadtree.

Before we introduce the algorithms for creating PM quadtrees, we need to address two important data issues. First, since we are dealing with lines and polygons, it is necessary to have a good data structure to store such information. For points, we have used the `Point` class which contains all the critical information we need, such as the coordinates and some essential operations to work on points. For lines and polygons, we could use the doubly connected edge list (DCEL) introduced in Chapter 3. However, the DCEL uses the concept of the half-edge to enforce the topological integrity between points, lines, and polygons. While effective for that purpose, DCEL does not explicitly give us information about edges per se, because each edge in our data will have two half-edges, which is inconvenient when we only want to store edges in PM quadtrees. For this reason, we extend the DCEL into a new class called the XDCEL (Listing 7.1, where X means extended), where we specifically define an `Edge` class that stores the from and to nodes, and the left and right polygons for each edge (lines 10–13), as can be found in any standard GIS textbook.[1] The `eq` function starting at line 14 determines whether two edges are equivalent (which they are if they have the same endpoints), regardless of the order of the points in an edge. The `extent` function returns the minimum and maximum X and Y coordinates of the edge as an instance of the `Extent` class (see below). The member function `is_endpoint` will become useful when it comes to checking whether several edges share the same point.

The XDCEL class itself (line 30) inherits from the DCEL class. Here we collect the half-edges, vertices, and faces from DCEL and then add a list of edges (line 35) as part of the entire data structure. The edges will be built using data from DCEL in the member function `build_xdcel`, which essentially traverses all the half-edges to figure out the from and to nodes and the left and right polygons for the edges.

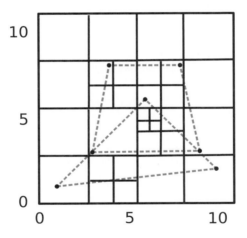

Figure 7.1 Partitioning of space for a PM1 quadtree

---

[1]However, it should be noted that an edge in DCEL is simply a straight line segment, while an edge in the strict sense of topological relationship in GIS is the chain or arc between two nodes that are the intersection of chains. In our DCEL context we count all vertices, at intersections or not, as the endpoints of edges. In a DCEL we may store redundant information for the multiple edges between two intersection nodes because these edges have exactly the same left and right polygon. But doing so also gives us the convenience of having all the vertices in our data so that we do not need to maintain a separate set of point data for the non-node vertices.

Listing 7.1: Extended doubly connected edge list (xdcel.py).

```python
import sys
sys.path.append('../geom')
from dcel import *
from extent import *

class Edge:
    """An edge in a topologically linked spatial data set"""
    def __init__(self, v1, v2, leftpoly, rightpoly):
        # Origin is defined as the vertex it points to (v2)
        self.fr = v1
        self.to = v2
        self.left = leftpoly
        self.right = rightpoly
    def __eq__(self, other):        # v1->v2 is eq. to v2->v1
        return (self.fr == other.fr and self.to == other.to)\
            or (self.fr == other.to and self.to == other.fr)
    def extent(self):
        xmin = min(self.fr.x, self.to.x)
        xmax = max(self.fr.x, self.to.x)
        ymin = min(self.fr.y, self.to.y)
        ymax = max(self.fr.y, self.to.y)
        return Extent(xmin, xmax, ymin, ymax)
    def is_endpoint(self, p):
        if p == self.fr or p==self.to:
            return True
        return False
    def __repr__(self):
        return "{0}->{1}".format(self.fr, self.to)

class Xdcel(Dcel):
    def __init__(self, D):
        Dcel.hedges = D.hedges
        Dcel.vertices = D.vertices
        Dcel.faces = D.faces
        self.edges=[]
        self.build_xdcel()
    def build_xdcel(self):
        """Build the Edges using DCEL info"""
        if not len(self.vertices) or not len(self.hedges):
            return
        for h in self.hedges:
            v1 = h.origin
            v2 = h.nextedge.origin
            lf = h.face
```

Indexing Lines and Polygons

```
45              rf = h.twin.face
46              e = Edge(v1, v2, lf, rf)
47              try:
48                  i = self.edges.index(e)
49              except ValueError:
50                  i = None
51              if i is None:
52                  self.edges.append(e)
```

The second data issue is the extent of each edge. We use the `Extent` class to organize the coordinate bounds of a geometric object (Listing 7.2). This is a generic class that can be used not only for the edges here, but also for polygons in the next section. This class contains essential data for defining an extent: the boundaries of the X and Y coordinates. An extent is also a bounding box or rectangle of the object. By overwriting the `__getitem__` function (line 12), an extent instance can be used as a list, with the indices 0, 1, 2, and 3 used to retrieve the four elements of the bounds. The other member functions of this class have names that should be self-explanatory. The member function `distance` returns the shortest distance between two extents by looking at the two nearest corners of the two extents.

Listing 7.2: Extent class (extent.py).

```
1  from math import sqrt
2  import sys
3  sys.path.append('../geom')
4  from point import *
5
6  class Extent():
7      def __init__(self, xmin=0, xmax=0, ymin=0, ymax=0):
8          self.xmin = xmin
9          self.xmax = xmax
10         self.ymin = ymin
11         self.ymax = ymax
12     def __getitem__(self, i):
13         if i==0: return self.xmin
14         if i==1: return self.xmax
15         if i==2: return self.ymin
16         if i==3: return self.ymax
17         return None
18     def __repr__(self):
19         return "[{0}, {1}, {2}, {3}]".format(
20             self.xmin, self.xmax ,self.ymin, self.ymax)
21     def __eq__(self, other):
22         return self.xmin==other.xmin and\
23             self.xmax==other.xmax and \
24             self.ymin==other.ymin and \
25             self.ymax==other.ymax
```

```
26      def touches(self, other):
27          return not (self.xmin>other.xmax or\
28                      self.xmax<other.xmin or\
29                      self.ymin>other.ymax or\
30                      self.ymax<other.ymin)
31      def contains(self, point):
32          return not (self.xmin>point.x or\
33                      self.xmax<point.x or\
34                      self.ymin>point.y or\
35                      self.ymax<point.y)
36      def getcenter(self):
37          return Point(self.xmin+(self.xmax-self.xmin)/2.0,
38                       self.ymin+(self.ymax-self.ymin)/2.0)
39      def getwidth(self):
40          return self.xmax-self.xmin
41      def getheight(self):
42          return self.ymax-self.ymin
43      def is_minimal(self):
44          return (self.xmax-self.xmin)<0.1 and\
45              (self.ymax-self.ymin)<0.1
46      def area(self):
47          return (self.xmax-self.xmin)*(self.ymax-self.ymin)
48      def intersect(self, other):
49          if not self.touches(other):
50              return 0
51          xmin = max(self.xmin, other.xmin)
52          xmax = min(self.xmax, other.xmax)
53          ymin = max(self.ymin, other.ymin)
54          ymax = min(self.ymax, other.ymax)
55          return (xmax-xmin)*(ymax-ymin)
56      def distance(self, other):
57          x1 = self.xmin+(self.xmax-self.xmin)/2.0
58          y1 = self.ymin+(self.ymax-self.ymin)/2.0
59          x2 = other.xmin+(other.xmax-other.xmin)/2.0
60          y2 = other.ymin+(other.ymax-other.ymin)/2.0
61          return sqrt((x1-x2)*(x1-x2)+(y1-y2)*(y1-y2))
```

7.1.1 PM1 quadtrees

A general procedure for creating a polygonal map quadtree is to create the point region quadtree first and then consider whether quadrants should be further partitioned. We first define the data structure that can be used to hold information for a node in a PM quadtree. This is the class called PMQuadTreeNode in Listing 7.3. Here we can observe the resemblances to the node structure we used in PR *k*-D trees which also has a member called center, which refers to the center of the quadrant represented by the node. Here, we explicitly use an `Extent` class to represent the bounds of the quadrant, and the same `Extent` class will be used later in this chapter. We could use the `Extent` class for the PR *k*-D tree too (see Exercise 1 at the end of this chapter). In the PM quadtree node, we package the four quadrants of the node together in a list called `quads`. As will be evident later in the code, such a change provides a great deal of convenience in implementing the algorithms. A node may contain a number of edges, and we store them in the list `edges`. If the region of the node contains a data point, it is stored as the `vertex` member. In PM quadtrees, we use a variable `type` to specify whether a node is a non-leaf node (a grey node), a node whose region contains a vertex or intersects with lines (black node), or a node that does not contain a vertex or intersect with any lines (white node). Evidently, black and white nodes are leaf nodes.

The first step in creating a PM1 quadtree is to create a PR quadtree. This is done using the function `split_by_points` (line 30), where the space is split in a recursive fashion. During the process of splitting the space, a key task in this algorithm is to compute the extents of the new quadrants using the bounds and center of the current extent (lines 38–47). With the next extents, we can create four new nodes (line 49). Then the subset of points that belongs to each smaller region is calculated (line 52). The subsets of points and the new nodes are then used to continuously split the space, until each quadrant contains no more than one point (lines 31 and 35). The function `search_pmquadtree` (line 59) implements an algorithm that continuously searches for the region containing the target point until a leaf node (not grey) is reached. Finally, the function `is_intersect` implements a simple algorithm that tests whether each of the four boundaries of an extent intersects with an edge. Here we utilize the algorithms discussed in Section 3.1 to test whether two line segments intersect.

Listing 7.3: Code common to all PM quadtrees (pmquadtree.py).

```
1   # Used by all PM quadtrees
2   import sys
3   sys.path.append('../geom')
4   from extent import *
5   from point import *
6   from linesegment import Segment
7   from intersection import test_intersect
8
9   BLACK = 2 # nodes with point or line
10  WHITE = 1 # nodes with no point or line intersecing
11  GREY  = 0 # intermediate nodes
12
13  class PMQuadTreeNode():
14      def __init__(self, point, extent,
15                   nw=None, ne=None, se=None, sw=None):
16          self.point = point # center
17          self.extent = extent
18          self.quads = [nw, ne, se, sw]
19          self.vertex = None
```

```
            self.edges = []
            self.type = GREY
    def __repr__(self):
        return str(self.point)
    def __getitem__(self, i):
        if i<4: return self.quads[i]
        return None
    def is_leaf(self):
        return sum([ q is None for q in self.quads])==4

def split_by_points(points, pmq):
    if len(points) == 1:
        pmq.vertex = points[0]
        pmq.type = BLACK
        return
    if len(points) ==0:
        pmq.type = WHITE
        return
    xmin = pmq.extent.xmin
    xmax = pmq.extent.xmax
    ymin = pmq.extent.ymin
    ymax = pmq.extent.ymax
    xmid = xmin + (xmax-xmin)/2.0
    ymid = ymin + (ymax-ymin)/2.0
    exts = [ Extent(xmin, xmid, ymid, ymax),  # nw
             Extent(xmid, xmax, ymid, ymax),  # ne
             Extent(xmid, xmax, ymin, ymid),  # se
             Extent(xmin, xmid, ymin, ymid)   # sw
    ]
    pmq.quads = [PMQuadTreeNode(exts[i].getcenter(),exts[i])
                 for i in range(4)]
    subpoints = [[], [], [],    []] # four empty points lists
    for p in points:
        for i in range(4):
            if exts[i].contains(p):
                subpoints[i].append(p)
    for i in range(4):
        split_by_points(subpoints[i], pmq.quads[i])

def search_pmquadtree(pmq, x, y):
    if pmq.type is not GREY:
        return pmq
    for q in pmq.quads:
        if q.extent.contains(Point(x, y)):
            return search_pmquadtree(q, x, y)
    return None
```

```
67  def is_intersect(extent, edge):
68      if not extent.touches(edge.extent()):
69          return False
70      # four corners clockwise
71      p1 = Point(extent.xmin, extent.ymin)
72      p2 = Point(extent.xmin, extent.ymax)
73      p3 = Point(extent.xmax, extent.ymax)
74      p4 = Point(extent.xmax, extent.ymin)
75      segs = [ Segment(0, p1, p2), Segment(1, p2, p3),
76               Segment(2, p3, p4), Segment(3, p4, p1) ]
77      s0 = Segment(4, edge.fr, edge.to)
78      for s in segs:
79          if test_intersect(s, s0):
80              return True
81      return False
```

The function `split_by_edges_pm1` in Listing 7.4 implements a recursive algorithm to finalize the building of a PM1 quadtree. The function takes two parameters: a list of the `Edge` instances and a PR quadtree node that contains all the vertices of the edges. The algorithm starts by figuring out the edges that intersect with the region extent of the input node (line 8), which could happen in two cases when an edge actually intersects the region bounds (line 9) or is completely contained within the region (line 12). When no such edges exist (line 14), we set the node type to be white to indicate that it is a leaf node without a point or edge and no further partitioning is required (therefore `return`). For a node whose region contains a vertex (line 17), we check if the edges intersecting this region are all from the same vertex inside the region (the `for` loop from line 19). If all edges in the region are indeed from the same vertex (line 22), we add the edges to the `edges` member of the node, set the type of the node to black, and stop the function call (`return`) on this node because there is no need to further partition the region. If the edges are from different vertices, we will continue to split the region.

When the node does not contain a vertex (line 27), we stop splitting the space only when there is one edge intersecting the region. In that case (line 28), we set the type to be black, add the edge to the `edges` member of the node, and stop the function call (`return`). Before we do any further processing, line 32 is necessary because we want to avoid situations where splitting will continue indefinitely, or the regions become infinitesimally small before the procedure stops. Such a situation occurs when two lines intersect a region and both lines are very close to each other and very close to one of the corners of the region. When these lines are not from the vertex in the region, this situation entails a very deep partitioning before we can separate them into two different quadrants.

Having checked all these special cases, we are at the point where we must physically split the region into four additional quadrants. There are two tasks that remain to be done. First, we need to figure out the extents of these smaller quadrants. We have done this before in the `split_by_points` function to create the initial PR quadtree (Listing 7.3). Here we use exactly the same strategy to get the new extents and create the new quadrants of the node (lines 38–47). Of course, we will only do this if the node is a leaf node of the PR quadtree (line 37). It will make no sense to split a non-leaf node (because it is already split – that is why it is called a non-leaf node). The second task is to decide which of the four new quadrants gets the vertex (line 52). After that, we make sure the current node does not contain a vertex (line 56), since one of its new child nodes will contain the point.

After all the above is done, we recursively call the same function four times, taking each of the new quadrants as a new node to be either assigned as a leaf node or split further (line 58 in Listing 7.4). Since we have considered all possible cases to decide whether a node should be split or not, the recursive call will stop successfully.

Listing 7.4: PM1 quadtree (pm1quadtree.py).

```
import sys
sys.path.append('../geom')
from xdcel import *
from pmquadtree import *

def split_by_edges_pm1(edges, pmq):
    subedges = []
    for e in edges:
        if is_intersect(pmq.extent, e):
            subedges.append(e)
        elif pmq.extent.contains(e.fr) and\
            pmq.extent.contains(e.to):
            subedges.append(e)
    if len(subedges) == 0:
        pmq.type = WHITE
        return
    if pmq.vertex is not None:
        is_same_source = True
        for e in subedges:
            if not e.is_endpoint(pmq.vertex):
                is_same_source = False
        if is_same_source:
            for e in subedges:
                pmq.edges.append(e)
            pmq.type = BLACK
            return
    else: # pmq does not contain a vertex
        if len(subedges) == 1:
            pmq.type = BLACK
            pmq.edges.append(subedges[0])
            return
    if pmq.extent.is_minimal():
        for e in subedges:
            pmq.edges.append(e)
        pmq.type = BLACK
        return
```

```
37        if pmq.is_leaf(): # now we split this if necessary:
38            xmin = pmq.extent.xmin
39            xmax = pmq.extent.xmax
40            ymin = pmq.extent.ymin
41            ymax = pmq.extent.ymax
42            xmid = xmin + (xmax-xmin)/2.0
43            ymid = ymin + (ymax-ymin)/2.0
44            exts = [ Extent(xmin, xmid, ymid, ymax),  # nw
45                     Extent(xmid, xmax, ymid, ymax),  # ne
46                     Extent(xmid, xmax, ymin, ymid),  # se
47                     Extent(xmin, xmid, ymin, ymid)   # sw
48                   ]
49            pmq.quads = [ PMQuadTreeNode(exts[i].getcenter(),
50                                          exts[i])
51                          for i in range(4) ]
52        if pmq.vertex:
53            for q in pmq.quads:
54                if q.extent.contains(pmq.vertex):
55                    q.vertex = pmq.vertex
56            pmq.vertex = None
57        for i in range(4):
58            split_by_edges_pm1(subedges, pmq.quads[i])
```

In order to test the PM1 algorithm, we will need the DCEL data. Here, we use the polygons shown in Figure 7.1, which is the result of the overlay operation described in Section 3.2. In practice, we can store a Python object to a file by serializing the binary information into a format that can be saved in a file. More specifically, we can add the following two lines to the end of the code in Listing 3.6. This will create an ASCII file called mydcel.pickle that stores all the information in the DCEL object D.

```
import pickle
pickle.dump(D, open('mydcel.pickle', 'w'))
```

Now we can use the file by unpickling or deserializing it. The following code segment illustrates how we restore the information in D (line 4) and then use that to create an XDCEL object XD (line 5). Lines 7–12 create the largest extent, which is used to construct the root node (line 14). Lines 15 and 16 create the PM1 quadtree and line 17 searches for the node at location (1, 1). Figure 7.2 shows the PM1 quadtree for the sample data.

```
1  import pickle
2  from pm1quadtree import *
3
4  D = pickle.load(open('mydcel.pickle'))
5  XD = Xdcel(D)
6
7  X = [v.x for v in D.vertices]
8  Y = [v.y for v in D.vertices]
```

```
9   xmin,xmax,ymin,ymax = min(X)-1, max(X)+1, min(Y)-1, max(Y)+1
10  maxmax = max(xmax,ymax)
11  xmax=ymax=maxmax
12  extent = Extent(xmin, xmax, ymin, ymax)
13
14  pmq = PMQuadTreeNode(extent.getcenter(), extent)
15  split_by_points(XD.vertices, pmq)
16  split_by_edges_pm1(XD.edges, pmq)
17  print search_pmquadtree(pmq, 1, 1)
```

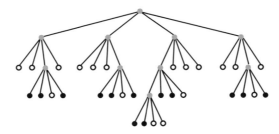

Figure 7.2 PM1 quadtree. Grey nodes represent intermediate quadrants that will be split, white nodes are those regions that do not contain a vertex or an edge, and black nodes refer to those regions that contain a vertex or intersect with edges

7.1.2 PM2 quadtrees

A drawback of PM1 quadtrees is their depth: the tree can be too deep because regions must be split if they intersect with lines from different vertices, and it therefore may require much storage. In the PM2 quadtree, we allow multiple edges in a region as long as they share the same vertex, even if the shared vertex is outside the region. In other words, a region is not split if all the line segments that it contains intersect at the same vertex outside the region. This could significantly reduce the number and depth of splits when constructing the tree. Figure 7.3 illustrates the case of a PM2 quadtree. Compared with the PM1 quadtree in Figure 7.1, it is noticeable that the quadrants in the lower half of the entire area do not need to be further split because for each quadrant the lines share a common vertex.

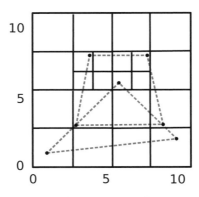

Figure 7.3 Partitioning of the space for a PM2 quadtree

The algorithm to construct a PM2 quadtree is shown in Listing 7.5. While this algorithm is similar to that for PM1 quadtrees, some parts of it are specifically designed for the PM2 tree. First, if a region only contains one edge (line 16), then we specify the node as a leaf (type black). Otherwise, if there are multiple lines (line 20) that share the same vertex (lines 23–31), regardless of where the vertex is located, we specify the node as a black leaf node (line 35). The `test` function uses the same XDCEL data as used for the PM1 quadtree.

Listing 7.5: PM2 quadtree (pm2quadtree.py).

```
1   from xdcel import *
2   from pmquadtree import *
3   import pickle
4
5   def split_by_edges_pm2(edges, pmq):
6       subedges = []
7       for e in edges:
8           if is_intersect(pmq.extent, e):
9               subedges.append(e)
10          elif pmq.extent.contains(e.fr) and\
11              pmq.extent.contains(e.to):
12              subedges.append(e)
13      if len(subedges) == 0:
14          pmq.type = WHITE
15          return
16      elif len(subedges) == 1:
17          pmq.type = BLACK
18          pmq.edges.append(subedges[0])
19          return
20      else:
21          p1,p2 = subedges[0].fr, subedges[0].to
22          common_vertex = None
23          if subedges[1].is_endpoint(p1):
24              common_vertex = p1
25          elif subedges[1].is_endpoint(p2):
26              common_vertex = p2
27          if common_vertex is not None:
28              for e in subedges[2:]:
29                  if not e.is_endpoint(common_vertex):
30                      common_vertex = None
31                      break
32          if common_vertex is not None:
33              for e in subedges:
34                  pmq.edges.append(e)
35              pmq.type = BLACK
36              return
```

```python
37        if pmq.extent.is_minimal():
38            for e in subedges:
39                pmq.edges.append(e)
40            pmq.type = BLACK
41            return
42        if pmq.is_leaf():
43            xmin = pmq.extent.xmin
44            xmax = pmq.extent.xmax
45            ymin = pmq.extent.ymin
46            ymax = pmq.extent.ymax
47            xmid = xmin + (xmax-xmin)/2.0
48            ymid = ymin + (ymax-ymin)/2.0
49            exts = [ Extent(xmin, xmid, ymid, ymax), # nw
50                     Extent(xmid, xmax, ymid, ymax), # ne
51                     Extent(xmid, xmax, ymin, ymid), # se
52                     Extent(xmin, xmid, ymin, ymid)  # sw
53                   ]
54            pmq.quads = [ PMQuadTreeNode(exts[i].getcenter(),
55                                         exts[i])
56                          for i in range(4) ]
57        if pmq.vertex:
58            for q in pmq.quads:
59                if q.extent.contains(pmq.vertex):
60                    q.vertex = pmq.vertex
61            pmq.vertex = None
62        for i in range(4):
63            split_by_edges_pm2(subedges, pmq.quads[i])
64
65  def test():
66      D = pickle.load(open('../data/mydcel.pickle'))
67      XD = Xdcel(D)
68
69      X = [v.x for v in D.vertices]
70      Y = [v.y for v in D.vertices]
71      xmin,xmax,ymin,ymax = min(X)-1, max(X)+1,\
72                            min(Y)-1, max(Y)+1
73      maxmax = max(xmax,ymax)
74      xmax=ymax=maxmax
75      extent = Extent(xmin, xmax, ymin, ymax)
76
77      pm2q = PMQuadTreeNode(extent.getcenter(), extent)
78      split_by_points(XD.vertices, pm2q)
79      split_by_edges_pm2(XD.edges, pm2q)
80      print search_pmquadtree(pm2q, 10, 10)
81
82  if __name__ == '__main__':
83      test()
```

7.1.3 PM3 quadtrees

Using the PM3 quadtree, we can further reduce the possibility of splitting regions because now we only partition a region if it contains more than one vertex. This makes the PM3 quadtree the same as a point region (PR) quadtree. Figure 7.4 shows an example of the split using the same data as before. Implementing the PM3 quadtree is straightforward because there is no further partitioning of the region. The only task we must complete here is to make sure the nodes contain information about the edges, as detailed in the `split_by_edges_pm3` function in Listing 7.6.

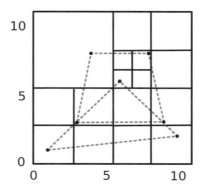

Figure 7.4 Partitioning of the space for PM3 quadtree

Listing 7.6: PM3 quadtree (pm3quadtree.py).

```
1  from xdcel import *
2  from pmquadtree import *
3  import pickle
4
5  def split_by_edges_pm3(edges, pmq):
6      subedges = []
7      for e in edges:
8          if is_intersect(pmq.extent, e):
9              subedges.append(e)
10         elif pmq.extent.contains(e.fr) and\
11            pmq.extent.contains(e.to):
12             subedges.append(e)
13     if not pmq.is_leaf():
14         for i in range(4):
15             split_by_edges_pm3(subedges, pmq.quads[i])
16         return
17     if len(subedges) == 0:
18         pmq.type = WHITE
19     return
```

```
20          else:
21              pmq.type = BLACK
22              for e in subedges:
23                  pmq.edges.append(e)
24              return
25
26      def test():
27          D = pickle.load(open('../data/mydcel.pickle'))
28          XD = Xdcel(D)
29
30          X = [v.x for v in D.vertices]
31          Y = [v.y for v in D.vertices]
32          xmin,xmax,ymin,ymax = min(X)-1, max(X)+1,\
33                                min(Y)-1, max(Y)+1
34          maxmax = max(xmax,ymax)
35          xmax=ymax=maxmax
36          extent = Extent(xmin, xmax, ymin, ymax)
37
38          pm3q = PMQuadTreeNode(extent.getcenter(), extent)
39          split_by_points(XD.vertices, pm3q)
40          split_by_edges_pm3(XD.edges, pm3q)
41          print search_pmquadtree(pm3q, 1, 1)
42
43      if __name__ == '__main__':
44          test()
```

All the PM quadtrees can be used for querying points and lines. For example, we can search for the nearest line to a point by going down the tree while excluding the nodes that are far away from the point. To search for polygons, however, we will need the information relating each edge to its adjacent polygons, which is stored in the XDCEL object. We do not detail these mechanisms here, but interested readers can find useful information in the literature discussed in the notes at the end of this chapter.

7.2 R-trees

PM quadtrees can be used to index lines, and indirectly the polygons that are associated with lines. R-trees, however, are specifically designed to index polygons, so that searching for polygons and other operations can be expedited. Interestingly, a key concept in R-trees is not the polygon itself but the minimal bounding rectangle (MBR) that can be used to represent the polygon. Figure 7.5 shows the MBRs for three polygons, *A*, *B*, and *C* (R3, R4, and R5, respectively). While the MBR does not contain the detailed geometric information of the polygon, it can be very helpful in determining the location of the polygon as well as helping reduce the computation cost of carrying out searches and operations such as overlay. For example, to find the polygons that share boundaries with a given polygon, we can compare the rectangles to determine if two polygons could possibly touch each other. It is quicker to compare two MBRs compared to potentially more complex polygons. Such comparison is of course not perfect – rectangles R3 and R5, for example, intersect but their polygons (*A* and *C*) do not. However, this can greatly narrow down the search to a very small number of polygons, and it will not take too much computation before we arrive at a final result. By abstracting polygons using their MBRs, we are now able to ignore a lot of detail related to complicated polygons and focus on the essential operations.

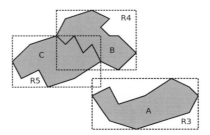

Figure 7.5 Minimal bounding rectangles for polygons

To index the MBRs in Figure 7.5 using an R-tree, we first define the node of a tree. Unlike our previous trees where each node only contains one object, here we allow each node to contain multiple objects, each called an entry. Before an R-tree is created, we must specify the maximum number of entries for a node, denoted by M. In our simple example, we set M to 2 to illustrate the concept. Another requirement of an R-tree is that each node must contain a minimum number of entries to avoid empty nodes. The minimum number of entries is typically defined as the ceiling integer of $M/2$, in our case 1. Finally, an R-tree is a balanced tree; all the original MBRs must be stored at the leaves of the tree and all the leaf nodes must have the same depth.

We are now ready to build an R-tree for the MBRs (Figure 7.6). We create a root node of the tree with two empty entries. We then start to insert the MBRs into the root. Suppose we insert the MBRs in the order R3, R4, and R5. The first two MBRs inserted will be R3 and R4, and the root can hold both of them. When the third MBR, R5, is inserted, however, the root is an overflowing node and cannot accommodate it, so the root node is split into two new nodes. Two things must happen to enable such a split. First, the two new nodes will be at a depth one level lower than the root. At this time, the two entries will need to be placed into the two new nodes, and it is necessary to determine which entry goes to which new node. The original R-tree algorithm dictates that the pair of rectangles that form the largest new bounding rectangle will be the worst combination and therefore each should be inserted into the two new nodes separately. Then, when there are more entries to be inserted into the new nodes, we always insert the MBR into a node so that minimal enlargement of the overall rectangle of the node is caused. In our simple example, since we only have two entries to split, R3 and R4 each go to a new node. Second, the linkages between the new nodes and the root need to be correctly built and the entries in the root need to adjust their bounding rectangles because they may now enclose multiple smaller MBRs after splitting. For each entry in the root node, it now has a new MBR that contains the combination of all the entries in one of the new nodes. We use a pointer associated with each entry to specify such a relationship between the entry and the child node. On the other hand, each child node needs a pointer that tells us which entry in the parent node encloses the child node. In Figure 7.6 we use arrows to illustrate these pointers. For the sake of convenience, we denote the two new entries in the root as R1 and R2, where R1 contains MBR R3 and R2 contains R4.

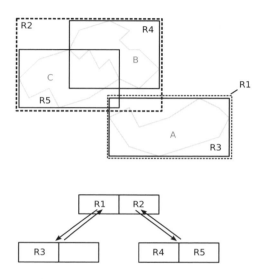

Figure 7.6 Nodes and entries of an R-tree

After splitting, we have room in the tree to accommodate new MBRs. We now insert the last MBR in our data, R5, into the tree. At the root level, we have two entries and we need to determine which entry the new MBR belongs to. We use the same principle as discussed above: we want to minimize the enlargement of bounding rectangles at the root level by adding the new MBR. Apparently, enlargement will be smallest if we insert R5 into the entry that contains R4. We go down the tree using the branch indicated by R2 that points to a node where there is room to insert an extra entry. We put R5 there and then update the bounding rectangle of R2 to reflect the addition of a new MBR. The R-tree is now finalized as shown in Figure 7.6.

To implement the algorithm for building an R-tree, we need a data structure for each entry and node. The `Entry` class in Listing 7.7 (line 3) has three members: the MBR of the entry as an instance of the `Extent` class, the child node whose entries are contained by this entry, and the node that contains this entry. The node class is called `RTreeNode` (line 12) and has four important members: the list of instances of the `Entry` class, the maximum number of entries in the node, the entry in the parent node that contains this node, and the extent of the node as an instance of the `Extent` class. In addition to the data members, a few useful member functions are also included in this class. The __getitem__ function overwrites the Python built-in function and makes a node instance iterable on the entries. The `is_leaf` and `is_root` functions are convenient for coding. Two important functions are `update` and `update_up`. The `update` function makes sure the extent of the node is a union of all the MBRs of its entries; it uses another function called `union_extent` that is defined at line 52. The `update_up` function is used to make sure that the change in this node is reflected all the way to the root.

Listing 7.7: Data structures of an entry and a node for an R-tree (rtree1.py).

```
1   from extent import *
2
3   class Entry():
4       def __init__(self, extent=None, child=None,
5                    parent=None, node=None):
6           self.MBR = extent
7           self.child = child      # a child node
8           self.node = node        # node containing this entry
9       def __repr__(self):
10          return str(self.MBR)
11
12  class RTreeNode():
13      def __init__(self, M, parent=None):
14          self.entries = []
15          self.M = M
16          self.parent = parent    # an entry in the parent node
17          self.extent = None
18      def __getitem__(self, i):
19          if i>=self.M or i>=len(self.entries):
20              return None
21          return self.entries[i]
22      def __repr__(self):
23          return str(self.extent)
24      def is_leaf(self):
25          for e in self.entries:
26              if e.child is not None:
27                  return False
28          return True
29      def is_root(self):
30          return self.parent is None
31      def update(self):
32          if not len(self.entries):
33              return
34          if self.entries[0] is not None:
35              self.extent = self.entries[0].MBR
36          for e in self.entries[1:]:
37              self.extent = union_extent(self.extent, e.MBR)
```

```
38      def update_up(self):
39          self.update()
40          if self.is_root():
41              return
42          self.parent.MBR = self.extent
43          self.parent.node.update_up()
44      def get_all_leaves(self, depth=0):
45          if not self.is_leaf():
46              for e in self.entries:
47                  e.child.get_all_leaves(depth+1)
48          else:
49              print depth, "-", self, self.entries
50              return
51
52  def union_extent(e1, e2):
53      xmin = min(e1.xmin, e2.xmin)
54      xmax = max(e1.xmax, e2.xmax)
55      ymin = min(e1.ymin, e2.ymin)
56      ymax = max(e1.ymax, e2.ymax)
57      return Extent(xmin, xmax, ymin, ymax)
```

Listing 7.8 includes the function `insert` that inserts a new MBR (e) into an R-tree. We first make sure we do not insert an extent that is already in the tree (line 5). If the node still has room to include a new MBR (line 9), we simply add the new entry to the entry list, update all the nodes related to this entry in the tree, and the insertion is complete. If the node is overflowing, we create two new nodes (lines 18 and 19), then in lines 20–34 we figure out the two MBRs that are farthest apart in the current node. These two MBRs will be used as seeds for each of the new nodes (lines 39 and 40). The `while` loop at line 43 continues to insert the remaining MBRs into each of the new nodes so that the increase in the total area of the MBRs in each node is minimized. Once the two new nodes are ready, we need to split the old node and using the `split` function.

The `split` function takes in the current node instance and the two new nodes just created. We first create two new empty entries using the extents from the new nodes (lines 106 and 107). If the node to be split is a root (line 108), we remove the entries in the root and we will reuse the node but update its entries. Otherwise, we use one of the new nodes to replace the current one and insert the other new node into the parent node using the `insert` function. This insertion may cause further splitting of nodes. Through this process, we make sure that all the MBRs are in the leaf nodes of the tree.

There are many ways to search using an R-tree. Here we give an example of finding the leaf node that contains a given extent with highest overlapping ratio. This is implemented in the function `search_rtree_extent` where the algorithm continuously goes down the tree following the entry that has the highest intersection area until a leaf node is reached. Later we will use this function to build an R-tree.

Listing 7.8: Insertion and node splitting for an R-tree (rtree2.py).

```python
from rtree1 import *

# e is an extent
def insert(node, e, child=None):
    for ent in node.entries:                    # already in tree
        if ent.MBR == e:
            return True
    entry = Entry(extent=e, child=child)  # create a new entry
    if len(node.entries) < node.M:              # there is room
        entry.node = node
        if entry.child is not None:
            entry.child.parent = entry
        node.entries.append(entry)
        node.update_up()
        return True
    M = node.M                      # overflowing node needs to be split
    m = math.ceil(float(M)/2)
    L1 = RTreeNode(M)
    L2 = RTreeNode(M)
    maxi, maxj = -1, -1
    maxdist = 0.0
    tmpentries = [ent for ent in node.entries]
    tmpentries.append(entry)
    M1 = len(tmpentries)
    # get the farthest apart MBRs as seeds
    for i in range(M1):
        for j in range(i+1, M1):
            d = tmpentries[i].MBR.distance(tmpentries[j].MBR)
            if d>maxdist:
                maxdist = d
                maxi = i
                maxj = j
    e1 = tmpentries[maxi]
    e2 = tmpentries[maxj]
    allexts = []                    # holds the rest of the MBRs
    for ext in tmpentries:
        if ext is not e1 and ext is not e2:
            allexts.append(ext)
    L1.entries.append(e1)
    L2.entries.append(e2)
    L1.update()
    L2.update()
    while len(allexts):
        numremained = len(allexts)
```

```
45              gotonode = None
46              if len(L1.entries) == m-numremained:
47                  gotonode = L1
48              elif len(L2.entries) == m-numremained:
49                  gotonode = L2
50              if gotonode is not None:
51                  while len(allexts):
52                      ext = allexts.pop()
53                      gotonode.entries.append(ext)
54              else:
55                  minarea = union_extent(L1.extent,L2.extent).area()
56                  minext = -1
57                  gotonode = None
58                  for i in range(len(allexts)):
59                      tmpext1 = union_extent(L1.extent,
60                                              allexts[i].MBR)
61                      tmparea1 = tmpext1.area() - L1.extent.area()
62                      tmpext2 = union_extent(L2.extent,
63                                              allexts[i].MBR)
64                      tmparea2 = tmpext2.area() - L2.extent.area()
65                      if min(tmparea1, tmparea2) > minarea:
66                          continue
67                      minext = i
68                      if tmparea1 < tmparea2:
69                          if tmparea1 < minarea:
70                              tmpgotonode = L1
71                              minarea = tmparea1
72                      elif tmparea2 < tmparea1:
73                          if tmparea2 < minarea:
74                              tmpgotonode = L2
75                              minarea = tmparea2
76                      else:
77                          minarea = tmparea1
78                          if L1.extent.area() < L2.extent.area():
79                              tmpgotonode = L1
80                          elif L2.extent.area() < L1.extent.area():
81                              tmpgotonode = L2
82                          else:
83                              if len(L1.entries) < len(L2.entries):
84                                  tmpgotonode = L1
85                              else:
86                                  tmpgotonode = L2
87                  if minext <> -1 and tmpgotonode is not None:
88                      ext = allexts.pop(minext)
89                      gotonode = tmpgotonode
90                      gotonode.entries.append(ext)
91              gotonode.update()
```

```python
92              for ent in L1.entries:
93                  ent.node = L1
94                  if ent.child is not None:
95                      ent.child.parent = ent
96              for ent in L2.entries:
97                  ent.node = L2
98                  if ent.child is not None:
99                      ent.child.parent = ent
100         split(node, L1, L2)
101         L1.update_up()
102         L2.update_up()
103         return True
104
105 def split(node, L1, L2):
106     entry1 = Entry(L1.extent)
107     entry2 = Entry(L2.extent)
108     if node.is_root():
109         node.entries = []
110         entry1.node = node
111         entry2.node = node
112         entry1.child = L1
113         entry2.child = L2
114         node.entries.append(entry1)
115         node.entries.append(entry2)
116         L1.parent = entry1
117         L2.parent = entry2
118         return
119     else:
120         entry1.node = L1
121         L1.parent = node.parent
122         L1.parent.child = L1
123         del node
124         insert(L1.parent.node, L2.extent, L2)
125         return
126
127 def search_rtree_extent(node, e):
128     if node.is_leaf():
129         return node
130     best_entry = None
131     intersect_area = -1
132     for ent in node.entries:
133         tmp_area = ent.MBR.intersect(e)
134         if tmp_area > intersect_area:
135             intersect_area = tmp_area
136             best_entry = ent
137     return search_rtree_extent(best_entry.child, e)
```

Figure 7.7 shows ten MBRs that we will index using an R-tree. It is interesting to note that the use of MBRs does not require polygons to be topologically correct. In other words, polygons can possibly intersect other polygons without having to create intersection points. R15 and R17 in this figure, for example, cross each other, which would require their polygons cross each other, or some MBRs contain polygons that have multiple, disconnected parts. While these situations are not common, they do exist in some spatial data sets and can be handled in the R-tree.

Listing 7.9 shows how to create an R-tree based on the data presented in Figure 7.7. The MBRs are stored in the list `MBRs` where the order of the rectangles is reshuffled. We convert the MBRs to a list of instances of `Extent` (line 18). The root node is created at line 16. All the extents are inserted into the root one by one (line 20) by always inserting into the node that has the highest overlap in area. Here we use a simple function called `get_all_leaves` to return all the leaf nodes along with their depth.

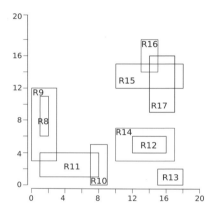

Figure 7.7 Minimal bounding rectangles used to create an R-tree

Listing 7.9: Using R-tree (use_rtree.py).

```
1   from rtree1 import *
2   from rtree2 import *
3
4   MBRs = [ [10,17,3,7],          # R14
5            [12,16,4,6],          # R12
6            [7,9,0,5],            # R10
7            [1,8,1,4],            # R11
8            [1,2,6,11],           # R8
9            [15,18,0,2],          # R13
10           [0,3,3,12],           # R9
11           [13,15,14,18],        # R16
12           [10,18,12,15],        # R15
13           [14,17,9,16] ]        # R17
```

```
14
15  M = 3
16  root = RTreeNode(M, None)
17
18  extents = [Extent(mbr[0], mbr[1], mbr[2], mbr[3])
19             for mbr in MBRs]
20  for e in extents:
21      n = search_rtree_extent(root, e)
22      insert(n, e)
23
24  root.get_all_leaves()
```

The following is the result of using the test code in Listing 7.9:

```
2 - [1, 9, 0, 5]    [[1, 8, 1, 4], [7, 9, 0, 5]]
2 - [0, 3, 3, 12]   [[0, 3, 3, 12], [1, 2, 6, 11]]
2 - [13, 17, 9, 18] [[13, 15, 14, 18], [14, 17, 9, 16]]
2 - [12, 18, 0, 6]  [[15, 18, 0, 2], [12, 16, 4, 6]]
2 - [10, 18, 3, 15] [[10, 17, 3, 7], [10, 18, 12, 15]]
```

It is clear that all the leaf nodes here come from the same depth, as required for an R-tree. Figure 7.8 shows how the MBRs are grouped and illustrates the corresponding R-tree.

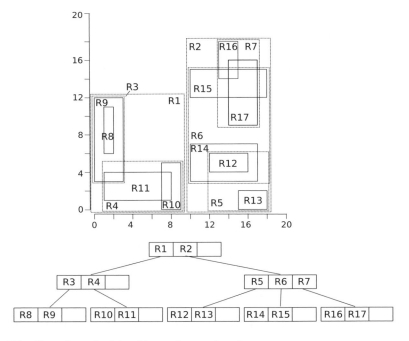

Figure 7.8 Grouping of minimal bounding rectangles

7.3 Notes

Samet and Webber (1985) first developed the idea of PM quadtrees. Among the three versions of PM quadtrees, PM1 may need excessive storage because of its edge representation requirements. PM3, however, may have too many points/lines in each node which could add computation time when the search process must test all the edges associated with each quadrant. PM2 appears to be a balanced comprise that is useful for polygonal maps.

The idea of quadtrees can be extended to higher dimensions. For example, octrees (Meagher, 1980, 1982) can be used to store points in a three-dimensional space, where the space is partitioned into eight blocks (octants) and the subdivision continues until certain criteria (e.g., each block contains more than one point) are satisfied. More information about these quadtrees can be found in Aluru (2005) as well as in the books by Samet (1990a,b, 2006).

The R-tree was invented by Guttman (1984) as a way to store and index polygon information. While our discussion of R-trees mainly relates to geospatial data, this idea can actually be used in a much wider context where objects must be quickly searched. Computer graphics, for example, is a major application domain of this type of indexing method. The key concepts used in R-trees, such as insertion and splitting, are similar to those in B-trees (Bayer and McCreight, 1972), a major indexing method that can be used for ordered objects (e.g., integers or anything that can sequenced). B-trees are beyond the scope of this book.

As with k-D trees, building R-trees can be expensive. In general, the time complexity of inserting n MBRs into a tree is $O(n)$. But searching an R-tree can be very efficient, with a running time of $O(M \log_M n)$, where M is the maximum number of entries in a node. We have not discussed other operations here. For example, it may be necessary to delete an MBR from an existing tree, making it necessary to restructure the tree accordingly. Details of these operations and more can of course be found in many books (Manolopoulos et al., 2006; Samet, 2006).

Through the discussion in this chapter, we can already understand some of the drawbacks of the R-trees. Addressing these drawbacks has been a dynamic field of research. For example, it is obvious that how the two seeds are determined when splitting a node can have a great impact on the overall structure of the tree and therefore the efficiency of using the tree. There have been other algorithms and variants of the R-tree. For example, the Hilbert R-tree uses the Hilbert space filling algorithm to partition space and group MBRs (Kamel and Faloutsos, 1994). Algorithms have also been proposed to address a major disadvantage of R-trees: the overlap between MBRs. The R$^+$-tree (Sellis et al., 1987), for example, is designed to avoid the overlap, and the R*-tree (Beckmann et al., 1990) to minimize the overlap.[2]

7.4 Exercises

1. Review the code for PR k-D trees and replace `xyrange` with the `Extent` class. Modify the PR k-D tree code accordingly.

2. Illustrate the PM2 and PM3 quadtrees for the space partitioned as shown in Figures 7.3 and 7.4.

3. Design and implement an algorithm to search an R-tree and find the MBRs that contain a given point.

---

[2]Note that the phrase R-tree is not read as "R minus tree" but just "R tree."

Part III

Spatial Analysis and Modeling

8

Interpolation

The overwhelming thrust of geospatial data in the age of "big data" gives us a false sense that we have obtained complete coverage of data observation around the world. In other words, we may erroneously think that we know everything about everywhere on the earth's surface. A proof: look at maps.google.com and choose the satellite view. We seem to be able to see *everywhere* on the map with a bird's-eye view (Figure 8.1). On the dark side of this, we have heard stories about how an advanced missile can hit almost anywhere on earth, with an accurate digital elevation model that guides the missile to hit the target.

Figure 8.1 Part of the Ohio State University campus shown in satellite view on Google Maps

To unveil the secrets behind the overwhelming explosion of data, we should understand a fundamental concept in geography and many other physical disciplines: scale. The term "scale" has different meanings in the geography literature. When we talk about maps, we refer to their scale as the ratio between the distance on the map and in reality. Sometimes we may refer to our work as a "large scale" project, meaning that the work is large in extent or broad in scope. In addition to these two meanings, the concept of scale has further implications when it is referred to in operational and measurement contexts. For example, a first class on atmospheric systems will make sure you understand the difference between weather and climate. We watch the weather channel but not the climate channel (if such a thing exists) because the former is about atmospheric conditions in a relatively shorter period of time (days) and smaller extent of space (regional), while climate generally refers to the dynamics in a longer (seasonal, annual, and longer) time and bigger area (global, for example). So it is clear that air systems have different operational scales and the differences in these scales require us to use different terminology to describe them. It is also clear that hurricanes and earthquakes operate on different scales.

The fourth aspect of scale is really what we are more interested in here: the measurement scale, referring to the smallest distinguishable part of an object being observed. For example, a photograph has a finite number of pixels, and each of those pixels must represent a certain size of the reality being captured in the photo. Two things limit what we can put on a photo here: the number of pixels that can be physically put on the photo and the smallest unit that can be recognized by the device used to take the photo. While both of these aspects will be improved because of technical breakthroughs, it is undeniable that there are possibly infinite pieces of information in reality that we want to squeeze into a finite number of pixels. Impossible, indeed. And here comes the compromise: we try to use enough samples of the reality as a substitute for the reality. This creates two sets of reality: knowns where we have the observations, and unknowns where we do not. What happens when we really want to know the measurements at locations where we do not have observations? We guess. In other words, we make estimations.

In different application domains, estimations are made in different ways and treat space differently. In remote sensing, as we have discussed so far, we make estimations by saying that all the areas in a pixel have the same measure. For other types of measurement, such as elevation, soil depth, and temperature, space is specifically realized as distances from the known locations to the unknown, as illustrated in Figure 8.2 where we need the information at locations with observations (solid circles) to estimate the measure at the location where observation is absent (the open circle).

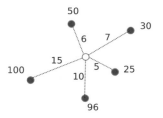

Figure 8.2 Interpolation. The solid dots represent locations with known measurements shown as the numbers near the dots. The open circle represents a location where no measurement is available and a value needs to be estimated. The distance between the unknown location and each of the known points is shown as a number near each dotted line

The technique used to estimate the values at locations where observations are not available is called interpolation. Cartographers and surveyors have known this technique for a long time. In the history of cartography and surveying, we aim to make sure that the known locations have accurate measures of their coordinates and elevation. This is traditionally achieved by establishing the geodetic networks around the world where each node is an observation point. From these networks, unknown elevations at other locations can be estimated. A reverse application is also possible: given some observed locations, we may want to estimate where a certain value would occur. A great use of this technique is the creation of topography maps where a set of contour lines is created to connect locations which we think should have the same values (Figure 8.3). In this chapter, we introduce two well established interpolation methods: inverse distance weighted interpolation and kriging. We will also discuss a simulation method that can be used to generate a surface that looks "real."

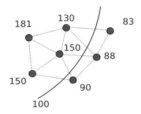

Figure 8.3 Interpolation of a contour that links points with a value of 100

8.1 Inverse distance weighted interpolation

Suppose we have a set of observations of a phenomenon, say elevation, and we want to use that to estimate the value of that phenomenon at a particular location. How do we go about that? Generally speaking, we can give each of the observed values a weight and then use that to obtain our estimate. Formally, given a set of n points where the values are known, we denote the measure at the ith point x_i by $z(x_i)$ and the weight for that location by λ_i. The unknown value at location x can be written as

$$z(x) = \frac{\sum_i \lambda_i z(x_i)}{\sum_i \lambda_i}.$$

We should note that the denominator sometimes is not necessary if we can make sure the sum of all the weights is 1 ($\sum_i \lambda_i = 1$).

A bigger question is how to assign weights to different locations. While we can do this in an arbitrary way, we may feel safer giving nearby locations higher weights. At the end, if there is anything that influences the value at a location, we would rather trust the locations nearby instead of the locations that are far away (so the elevation or temperature in California may not be as effective in predicting the value in Columbus, Ohio as those around central Ohio). Such a belief is actually captured in the first law of geography proposed by Tobler. Though many researchers have challenged the status of Tobler's statement as a law, it is a guideline we can use. It states that "everything is related to everything else, but near things are more related than distant things."

The inverse distance weighted (IDW) interpolation method applies exactly what the first law of geography advocates because it gives nearby points higher weights than distant points. More specifically, the weight given to a location with a known observation decreases with distance from the location to be estimated:

$$\lambda_i = \frac{1}{d_i^b},$$

where d_i is the distance from the ith observation to the unknown location, and b is a constant.

The code in Listing 8.1 implements the idea of the IDW method. The function IDW takes two arguments: the known data Z and the power b for calculating the interpolation. Here, Z is a list where each element is another list containing four elements: the X and Y coordinates, the value at the location, and the distance from this point to the unknown point (see below, where we use this for testing).

Listing 8.1: Inverse distance weighted interpolation (idw.py).

```
1  def IDW(Z, b):
2      """
3      Inverse distance weighted interpolation.
4      Input
5        Z: a list of lists where each element list contains
6           four values: X, Y, Value, and Distance to target
7           point. Z can also be a NumPy 2-D array.
8        b: power of distance
9      Output
10       Estimated value at the target location.
11     """
12     zw = 0.0                 # sum of weighted z
13     sw = 0.0                 # sum of weights
14     N = len(Z)               # number of points in the data
15     for i in range(N):
16         d = Z[i][3]
17         if d == 0:
18             return Z[i][2]
19         w = 1.0/d**b
20         sw += w
21         zw += w*Z[i][2]
22     return zw/sw
```

We use 100 elevation points in northeast Columbus, OH (Figure 8.4) to demonstrate the use of inverse distance weighting. The same data set will also be used in the next section when we discuss kriging. To test the IDW method, we first need to load the data points from a text file. Since we will do this kind of data reading many times in this chapter, we write a specific function called read_data in Listing 8.2. This function simply reads every line into a list (line 12) and then splits each line into elements using the Python string processing function split. This is done in list comprehension and the result is a list of lists (line 13). However, the items in the list are strings and we need to convert them into floating point numbers (line 14).

Interpolation

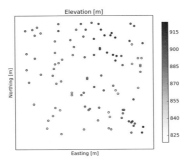

Figure 8.4 A sample data set of elevation points. The unit of length is the meter

---

Listing 8.2: Reading sample point data from a text file (read_data.py)

```
1  def read_data(fname):
2      """
3      Reads in data from a file. Each line in the file must have
4      three columns: X, Y, and Value.
5      Input
6        fname: name of and path to the file
7      Output
8        x3: list of lists with a dimension of 3 x n
9            Each inner list has 3 elements: X, Y, and Value
10     """
11     f = open(fname, 'r')
12     x1 = f.readlines()
13     x2 = [x.strip().split() for x in x1]
14     x3 = [[float(x[0]),float(x[1]),float(x[2])] for x in x2]
15     return x3
```

Another important task is to prepare data for interpolation. We use the function in Listing 8.3 to create a list of data for the `IDW` function. We calculate the distance between every sample point in the data to our unknown point, which can be done in one line (line 5) using Python's list comprehension. Line 7 organizes the necessary information from the data into a single list so that each element in the list relates to a point and contains the coordinates of a data point (X and Y), observed data, and distance to the location with unknown value. Afterwards, we sort the new data structure using the distance column (line 8) and pick the ten points that are nearest to the unknown point (line 9). In addition to returning the N points, the function also returns the mean value of the data which will be used later in this chapter.

Listing 8.3: Data preparation for interpolation (prepare_interpolation_data.py).

```
1  from math import sqrt
2  def prepare_interpolation_data(x, Z, N=10):
3      vals = [z[2] for z in Z]
4      mu = sum(vals)/len(vals)
```

```
5        dist = [sqrt((z[0]-x.x)**2 + (z[1]-x.y)**2) for z in Z]
6        Z1 = [(Z[i][0], Z[i][1], Z[i][2], dist[i])
7               for i in range(len(dist))]
8        Z1.sort(key=lambda Z1: Z1[3])
9        Z1 = Z1[:N]
10       return Z1, mu
```

We put the above code together to test the `IDW` function in Listing 8.4. We use an arbitrary unknown point at (337000, 4440911) in line 11 for testing purposes.

Listing 8.4: Testing inverse distance weighted interpolation (test_idw.py).

```
1   import sys
2   sys.path.append('../geom')
3   from point import *
4   from idw import *
5   from read_data import *
6   from math import sqrt
7   from prepare_interpolation_data import *
8
9   Z = read_data('../data/necoldem.dat')
10
11  x = Point(337000, 4440911)
12
13  N = 10
14
15  Z1 = prepare_interpolation_data(x, Z, N)[0]
16
17  print 'power=0.0:', IDW(Z1, 0)
18  print 'power=0.5:', IDW(Z1, 0.5)
19  print 'power=1.0:', IDW(Z1, 1.0)
20  print 'power=1.5:', IDW(Z1, 1.5)
21  print 'power=2.0:', IDW(Z1, 2.0)
```

Testing four different power values, the estimated values at the point are as follows:

```
power=0.0:  858.8
power=0.5:  859.268444166
power=1.0:  859.737719755
power=1.5:  859.951544335
power=2.0:  859.998590795
```

It is obvious that different values of b (power of distance) will yield different results. But which one should we use? The equation for calculating inverse distance weights is simple but not specific because we do not really know the value of b for the actual calculation. So how should we determine the value of b? It turns out that we will not know it until further investigation is done. In general, though, the higher the power of distance, the higher the weight assigned to nearby locations. When b is zero, all points have an equal weight of 1 and therefore we simply take the average of all values from the nearby locations as the estimate of the unknown value. A non-zero value of b gives distance the power to control the interpolation result. Figure 8.5 shows the impacts of different distance power on the resulting weights. To settle on an optimal b value, we will have to try out a few to find which one gives the best performance; this is typically measured using the root mean squared error (RMSE),

$$\text{RMSE} = \left(\frac{1}{n} \sum_j (p_j - o_j)^2 \right)^{1/2},$$

where p_j is the estimated value at location j ($1 \leq j \leq n$) and o_j is the observed value at location j. A traditional way to do such validation is to hold out a subset of the data so that they do not participate in interpolation. Instead, the subset of data that are held out are used to evaluate the interpolation results.

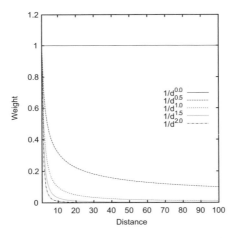

Figure 8.5 The impact of the power on inverse distance weights

There are two other things we can do using the IDW algorithm: we can test the impact of the power of distance and we can use the best power value to actually generate a surface. We will do the former here and leave the latter to the next section.

To test the impact of the power of distance, we will hold one of the known points out and estimate its value using the IDW function. We then do this for all the known points using a given power value and calculate the RMSE for that value. After a simple comparison, the optimal value of b should emerge.

Listing 8.5: Cross-validation of IDW (idw_cross_validation.py).

```
1  import sys
2  sys.path.append('../geom')
3  from point import *
4  from idw import *
```

```
5   from read_data import *
6   from math import sqrt
7   from prepare_interpolation_data import *
8
9   Z = read_data('../data/necoldem.dat')
10  N = len(Z)
11  numNeighbors = 10
12  mask = [True for i in range(N)]
13  powers = [0, 0.5, 1, 1.5, 2, 2.5, 3, 3.5, 4, 4.5, 5]
14  test_results = []
15  for i in range(N):
16      mask[i] = False
17      x = Point(Z[i][0], Z[i][1])
18      P = [ Z[j] for j in range(N) if mask[j] == True]
19      P1 = prepare_interpolation_data(x, P, numNeighbors)[0]
20      diff = []
21      for n in powers:
22          zz = IDW(P1, n)
23          diff.append(zz-Z[i][2])
24      test_results.append(diff)
25      mask[i] = True
26
27  for i in range(len(powers)):
28      rmse = sqrt(sum([r[i]**2 for r in test_results])/
29                  len(test_results))
30      print rmse, '[', powers[i], ']'
```

The code in Listing 8.5 can be used to cross-validate the distance power b. Here a mask array (line 16) is used to go through each of the observation points. It clearly shows that the RMSE decreases with the distance power until it reaches its smallest value when $b = 1.5$.

```
14.8683220304 [ 0 ]
14.1634839958 [ 0.5 ]
13.5663662616 [ 1 ]
13.2774718585 [ 1.5 ]
13.3066095177 [ 2 ]
13.5193354447 [ 2.5 ]
13.7913037092 [ 3 ]
14.060307198 [ 3.5 ]
14.3081392974 [ 4 ]
14.5338050919 [ 4.5 ]
14.7398249928 [ 5 ]
```

8.2 Kriging

The inverse distance weighted method introduced in the previous section can provide an estimate of the value of the phenomenon of interest at a given location. However, we have observed the arbitrary nature of IDW, especially when it comes to choosing the right value for the distance power b. Even though we can use the cross-validation method to obtain an optimal value that gives the smallest error, we might desire a method that would give us some hint about theoretically how the interpolation performs. The method called kriging addresses that issue. In the literature it is even called the optimal interpolation method, mainly because it uses a theoretical error model.

8.2.1 Semivariance

The key idea of kriging is to link the variation of observed values to their distance. Specifically, we can define a special quantity called semivariance as

$$\gamma(h) = \frac{1}{2} E([z(x+h) - z(x)]^2)$$

$$= \frac{1}{2N} \sum_{i=1}^{N} [z(x_i + h) - z(x_i)]^2,$$

where x is a location and $x + h$ is another location at a distance h from x, $z(x)$ and $z(x + h)$ are the values at these two locations, E denotes the expectation or mean value (in this case, of the squared difference), N is the number of observation pairs where each pair contains two locations with a distance of h, and (x_i) is the ith location in the N pairs. In the kriging literature, h, is often called the lag. We call $\gamma(h)$ the *semi*variance because its value is half of the expected value of the squared difference. Figure 8.6 shows the semivariance of a data set that will be used later in this chapter. We note the overall trend of $\gamma(h)$ as the lag value (h) increases.

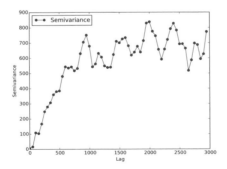

Figure 8.6 Empirical semivariance computed using a sample data set

To compute $\gamma(h)$, we need to find the pairs of observations that are a distance h apart. However, in reality, unless the observations are sampled systematically so that the distances between them are controlled, it is rare to have pairs of observations that have the exact distance h. To address this issue, we can utilize a set of distance bins such that the pairs that fall within the distance bin will be used to compute the semivariance of the average distance of that bin. We will use our sample point data to illustrate how to calculate the semivariance of a given data set.

Listing 8.6 is a Python program that can be used to return a two-dimensional array containing the semivariance at each lag point.[1] The `semivar` function takes three inputs, the data organized as a two-dimensional data set where each row contains the X and Y coordinates and the measure, the distances in the lags, and the size of the distance bins. Here, we use a few very useful features in the popular Python module for numerical analysis called NumPy (see Appendix A for some basics). The NumPy module makes operations on arrays and matrices extremely simple, not in terms of the actual computation, but of the expressive power for writing the code. We will see more of these features when we move on to other topics, but we have a good example here in line 30 that computes the distance matrix for all the pairs of the observations in the input data. The output of this line is an $n \times n$ matrix where n is the number of points in the data. For each lag point, we find all the pairwise distances in the data that fall into each of the bins (line 35) and put the squared difference in the corresponding observations into an empty list (line 36). We then calculate the mean of all those pairs (line 40) to get the semivariance at a given lag distance. Finally, we pair the lag distance and the associated semivariance (line 44) and return the values.

Listing 8.6: Computing semivariance (semivariance.py).

```
1  import numpy as np
2  from math import sqrt
3
4  def distance(a, b):
5      """
6      Computes distance between points a and b
7      Input
8        a: a list of [X, Y]
9        b: a list of [X, Y]
10     Output
11       Distance between a and b
12     """
13     d = (a[0]-b[0])**2 + (a[1]-b[1])**2
14     return sqrt(d)
15
16 def semivar(z, lags, hh):
17     """
18     Calculates empirical semivariance from data
19     Input:
20       z - a list or 2-D NumPy array,
21           where each element has X, Y, Value
22       lags - distance bins in 1-D array
23       hh - half of the bin size in distance
```

---

[1] The Python code for kriging, especially for semivariance, covariance, and simple kriging, is adopted and modified from https://github.com/cjohnson318/geostatsmodels.

```
24          Output:
25            A 2-D array of [ [h, gamma(h)], ...]
26          """
27          semivariance = []
28          N = len(z)
29          D = [[distance(z[i][0:2], z[j][0:2])
30                  for i in range(N)] for j in range(N)]
31          for h in lags:
32              gammas = []
33              for i in range(N):
34                  for j in range(N):
35                      if D[i][j] >= h-hh and D[i][j]<=h+hh:
36                          gammas.append((z[i][2]-z[j][2])**2)
37              if len(gammas)==0:
38                  gamma = 0
39              else:
40                  gamma = np.sum(gammas) / (len(gammas)*2.0)
41              semivariance.append(gamma)
42          semivariance = [ [lags[i], semivariance[i]]
43                              for i in range(len(lags))
44                              if semivariance[i]>0 ]
45          return np.array(semivariance).T
```

The following listing is the code that tests the above semivariance program using our sample data. The distance on each side of our sample area is 2,970 meters. We set the lag distance to range from 0 to 3,000 in steps of 50. In the output produced by the function call in line 10, a semivariance value is given for each lag distance. Figure 8.7 plots the semivariances against lag distances in the curve with dots. This is the empirical semivariance obtained from the data. Evidently, if we change the size of the bins we will get a different set of semivariance values. However, the general patterns should hold. It should be mentioned that the term "semivariance" is often used interchangeably with "semivariogram," which refers to the plot (hence gram or graph) in Figure 8.7. While the numbers are useful, the semivariogram often speaks more directly about the relationship between semivariance and lags.

Listing 8.7: Testing semivariance.py using the sample data.

```
1   import sys
2   sys.path.append('../geom')
3   from point import *
4   import numpy as np
5   from semivariance import *
6
7   Z = read_data('../data/necoldem.dat')
8   hh = 50
9   lags = np.arange(0, 3000, hh)
10  gamma = semivar(Z, lags, hh)
```

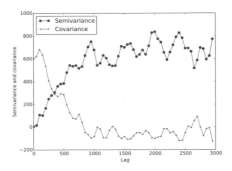

Figure 8.7 Empirical and modeled semivariance for the sample data

8.2.2 Modeling semivariance

We will use the semivariance to predict unknown values at locations that do not have observations. The empirical semivariance we have obtained so far is not ideal for this purpose. This is because the empirical semivariance will differ if we alter the way lags are used in the calculation and, perhaps more importantly, the empirical curve is "irregular," which makes it hard for us to obtain a $\gamma(h)$ value when it is not in the data. We need a theoretical curve. In other words, we need a mathematical function that reflects what we believe should be happening in reality and that can give us a semivariance value at any lag distance.

The empirical curve actually gives us some hints about what the theoretical curve should look like. First of all, the curve should start from a value when h is zero. We call this value a *nugget*. In the theory of geostatistics, a discipline developed around the concept of kriging, the nugget is the variation of observations even when the lag distance is zero – where two observations established side by side in the field have different values, the error is very likely to be caused by the instrument. Then we can observe that the semivariance seems to reach a certain value before it levels off. This value is called the *sill*, and the distance when the sill is reached is called the *range*. So in general we need to determine these three values: nugget, sill, and range. The fact of having the sill and range in many empirical cases shows how distance affects the variation of observed values: the longer the distance, the higher the variation, and hence the more dissimilar the phenomenon (as measured by its values).

Before we discuss how to actually find a theoretical model to fit the data, let us explore further the idea of similarity and dissimilarity. As mentioned above, the semivariance provides a way to measure the similarity between measures according to how far apart they are, and here we introduce another quantity to measure such similarity. This is the covariance between the pairs:

$$C(h) = E[(z(x) - \mu)(z(x+h) - \mu)]$$

$$= \frac{1}{N}\sum_{i=1}^{N} z(x_i) z(x_i + h) - \mu^2,$$

where μ is the mean observation value at lag distance h. In a very similar fashion to the `semivariance.py` program, we can write a new Python program to compute the covariance given the input data, bins of lags, and the length of the lag (Listing 8.8). For our sample data, the empirical covariance is shown as the curve with small dots having a general decreasing trend with lag distance (Figure 8.7).

Listing 8.8: Computing covariance (covariance.py).

```
import numpy as np
from semivariance import distance

def covar(z, lags, hh):
    """
    Calculates empirical covariance from data
    Input:
      z    - a list where each element is a list [x, y, data]
      lags - distance bins in 1-D array
      hh   - half of the bin size in distance
    Output:
      A 2-D array of [ [h, C(h)], ...]
    """
    covariance = []
    N = len(z)
    D = [ [distance(z[i][0:2],z[j][0:2])
              for i in range(N)] for j in range(N)]
    for h in lags:
        C = []
        mu = 0
        for i in range(N):
            for j in range(N):
                if D[i][j] >= h-hh and D[i][j]<=h+hh:
                    C.append(z[i][2]*z[j][2])
                    mu += z[i][2] + z[j][ 2]
        if len(C)==0:
            Ch = 0
        else:
            mu = mu/(2*len(C))
            Ch = np.sum(C) / len(C) - mu*mu
        covariance.append(Ch)
    covariance = [ [lags[i], covariance[i]]
                    for i in range(len(lags))]
    return np.array(covariance).T
```

It is obvious from Figure 8.7 that semivariance and covariance have opposite trends. Actually, if we assume that the phenomenon has mean and variance that are spatially constant, we can have the following relationships between these two quantities:

$$C(0) = E[(z(x)-\mu)^2] = \text{var}\,[z(x)],$$

and
$$\gamma(h) = C(0) - C(h),$$
where var denotes the variance (in this case, of the data $z(x)$).

As shown in Figure 8.7, when h is large enough (larger than the range distance, for example), $\gamma(h)$ tends to reach the sill and, on the other hand, $C(h)$ is zero or very close to zero. Therefore, we can set the sill to $C(0)$, which is the variance of the data as the first equation above shows. Now the question is: how can we obtain the values of the nugget and range? It is simple to get the nugget, which is the value of the semivariance at distance zero. Now we know how to estimate sill and nugget, and the parameter that is still unknown is the range. We will come back to this issue later. For now, we assume the nugget, sill, and range are given and we denote them as c_0, c, and a, respectively.

There are many different models that can be used to fit the data based on these three values. Among them, the linear model is the simplest that reaches the sill with a constant slope:

$$\gamma(h) = \begin{cases} c_0 + c\left(\frac{h}{a}\right) & 0 \leq h \leq a \\ c_0 + c & h > a \\ 0 & h = 0. \end{cases}$$

The spherical model is arguably the most commonly used and is written as

$$\gamma(h) = \begin{cases} c_0 + c\left[\frac{3h}{2a} - \frac{1}{2}\left(\frac{h}{a}\right)^3\right] & 0 \leq h \leq a \\ c_0 + c & h > a \\ 0 & h = 0. \end{cases}$$

The spherical model reaches the sill at the range distance. However, unlike the linear model, the spherical model reaches the sill with a decreasing slope as the lag distance increases. The exponential model is formulated as

$$\gamma(h) = \begin{cases} c_0 + c\left[1 - e^{\left(\frac{h}{a}\right)}\right] & h > 0 \\ 0 & h = 0, \end{cases}$$

which has a similar trend to the spherical model, but the former only reaches 95 percent of the sill value at the range distance. The Gaussian model has a sigmoid shape and does not reach the sill at the range distance:

$$\gamma(h) = \begin{cases} c_0 + c\left[1 - e^{\left(\frac{h^2}{a^2}\right)}\right] & h > 0 \\ 0 & h = 0. \end{cases}$$

Finally, we could also use a power function to model semivariograms as

$$\gamma(h) = \begin{cases} c_0 + c\,|h|^\lambda & h > 0 \\ 0 & h = 0, \end{cases}$$

where λ is a parameter between 0 and 2 that can be fitted using the data.

The Python code in Listing 8.9 completes the fitting of a semivariogram in the function called `fitsemivariogram` that takes three input variables: data (`z`), empirical semivariance (`s`), and one of the four models defined in the program. The format for defining these models can be easily extended to any other models requiring the lag (`h`) and the nugget (`c0`), sill (`c`), and range (`a`) parameters. Here we use a special feature in NumPy called function vectorization that is declared using the `@np.vectorize` command. Vectorization in Python allows the function to take arrays as input and to return arrays as output. We will take a close look at this below.

In Listing 8.9, line 56 computes the variance of the data, which, as discussed above, is equivalent to the value of the sill. Lines 57–60 give us the value of nugget. Then the program uses the minimum and maximum values of the range in line 61, assuming the largest range will not exceed the upper bound of the lag distance in the empirical semivariance. It is possible in reality for the actual range to be outside the largest lag distance in the data, but then we will have no data to fit the model. In this sense, while we are trying to create an automatic process to compute parameters for kriging, preparing the data takes careful thought. As we have seen so far, we will make sure we have the lag distance of zero ready in the data, otherwise we will not be able to calculate the nugget value.

We use a brute-force approach to estimate the value of the range (A) for each model: we test a large number of range values and the one that returns the least error wins. After the calculation of the empirical semivariance, we know the should-be semivariance values at the lag points so we can use those to evaluate how our model does. In the code, line 62 creates a number of ranges to test. By default we use 200 ranges. For each range, we calculate the mean squared difference between the estimated measure at all lag distances and the known observations (line 64). Let us pay attention to how the code works in line 64: because each of the models declared is vectorized, when an array is passed to the model function (here the array is `s[0]`, which contains all the lag distances in our data), it uses each of the elements in the array and returns an array of the same size, which is then used to get the squared difference of each of the lag distances. The distances form an array, and we will finally get the mean squared difference. The NumPy function `np.mean` calculates the mean error for each range tested. After testing all the potential range values, the one that yields the smallest error will be chosen as the range value (a) for the model (line 65). The function `fitsemivariogram` returns a lambda function (see Appendix A.3) that only takes one input (since all the other values, c_0, c, and a are calibrated).

Listing 8.9: Semivariogram fitting (fitsemivariance.py).

```python
import numpy as np

@np.vectorize
def spherical(h, c0, c, a):
    """
    Input
      h: distance
      c0: sill
      c: nugget
      a: range
    Output
      Theoretical semivariogram at distance h
    """
    if h<=a:
        return c0 + c*(3.0*h/(2.0*a) - ((h/a)**3.0)/2.0)
    else:
        return c0 + c

@np.vectorize
def gaussian(h, c0, c, a):
    """
    Same as spherical
    """
    return c0 + c*(1-np.exp(-h*h/((a)**2)))

@np.vectorize
def exponential(h, c0, c, a):
    """
    Same as spherical
    """
    return c0 + c*(1-np.exp(-h/a))

@np.vectorize
def linear(h, c0, c, a):
    """
    Same as spherical
    """
    if h<=a:
        return c0 + c*(h/a)
    else:
        return c0 + c

def fitsemivariogram(z, s, model, numranges=200):
    """
    Fits a theoretical semivariance model.
    Input
      z: data, NumPy 2D array, each row has (X, Y, Value)
      s: empirical semivariances
```

```
            model: one of the semivariance models: spherical,
                Gaussian, exponential, and linear
        Output
          A lambda function that serves as a fitted model of
          semivariogram. This function will require one parameter
          (distance).
        """
        c = np.var(z[:,2])          # c, sill
        if s[0][0] is not 0.0:      # c0, nugget
            c0 = 0.0
        else:
            c0 = s[0][1]
        minrange, maxrange = s[0][1], s[0][-1]
        ranges = np.linspace(minrange, maxrange, numranges)
        errs = [np.mean((s[1] - model(s[0], c0, c, r))**2)
                for r in ranges]
        a = ranges[errs.index(min(errs))]  # optimal range
        return lambda h: model(h, c0, c, a)
```

Continuing the code in Listing 8.7, we use the code in Listing 8.10 to fit the semivariogram of our sample data. Here, line 18 specifies the need to fit a spherical model using the data (Z) and the empirical semivariance (gamma). The last input for the fitsemivariogram function is the name of the function that is used to implement one of the models. In Python, each function name is also a variable name which can be directly used as an input argument for another function. Here we also fit other theoretical models: linear and Gaussian in lines 19 and 20, respectively. The rest of the program uses another powerful Python module called pylab that can create effective visualizations. The code itself here is straightforward to understand and to extend. It should be noted that the vectorization feature of functions is once again used here. For example, line 24 plots a list of data where the first set of elements are lag distances that are related to the horizontal axis and the second are computed using the fitted semivariogram for each of the lag distances.

Listing 8.10: Fitting semivariance using the sample data (test_fitsemivariance.py).

```
From pylab import *
import numpy as np
import sys
sys.path.append('../geom')
from point import *
from semivariance import *
from covariance import *
from read_data import *
from fitsemivariance import *

Z = read_data('../data/necoldem.dat')

hh = 50
lags = range(0, 3000, hh)
```

```
15   gamma = semivar(Z, lags, hh)
16   covariance = covar(Z, lags, hh)
17   Z1 = np.array(Z)
18   svs = fitsemivariogram(Z1, gamma, spherical)
19   svl = fitsemivariogram(Z1, gamma, linear)
20   svg = fitsemivariogram(Z1, gamma, gaussian)
21   sve = fitsemivariogram(Z1, gamma, exponential)
22
23   p1, = plot(gamma[0], gamma[1], 'o')
24   p2, = plot(gamma[0], svs(gamma[0]), color='grey', lw=2)
25   p3, = plot(gamma[0], svl(gamma[0]), color='grey', lw=2,
26              linestyle="--")
27   p4, = plot(gamma[0], svg(gamma[0]), color='grey', lw=2,
28              linestyle="-.")
29   p5, = plot(gamma[0], sve(gamma[0]), color='grey', lw=2,
30              linestyle=":")
31   models = ["Empirical", "Spherical", "Linear",
32             "Gaussian", "exponential"]
33   l1 = legend([p1,p2,p3,p4, p5], models, loc='lower right')
34   ylabel('Semivariance')
35   xlabel('Lag (m)')
36   savefig('semivariogram_data_model.eps',fmt='eps')
37   show()
```

Figure 8.8 shows the difference between the four fitted models using the code in Listing 8.10 (the line outputting an EPS file is omitted here). By visual examination, we can see that the spherical model actually provides a very reasonable set of estimates that are close to the data.

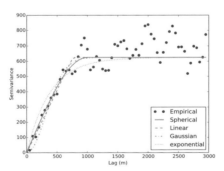

Figure 8.8 Modeled semivariograms using three models

8.2.3 Ordinary kriging

We now turn to the use of the semivariogram in kriging for interpolation. Similar to the inverse distance weighted interpolation method, in kriging we also calculate the unknown value at a location using a set of weights on some nearby values,

$$\hat{z}(x_0) = \sum_i \lambda_i z(x_i),$$

and we assume $\sum_i \lambda_i = 1$. A very important assumption in kriging is the stationarity hypothesis. Under this hypothesis, we assume that the mean and variance of the values are not correlated with their location, and formally we have $E[\hat{z}(x_0)] = E[z(x_0)] = E[z(x_i)]$ and $\text{var}[\hat{z}(x_0)] = \text{var}[z(x_0)] = \text{var}[z(x_i)]$, where x_i is any of the locations with known values in our data.

We use matrices to simplify the discussion of kriging. We will omit the notation x_0 for z and $\hat{z}$ in our discussion when there is no confusion. We estimate the value at location x_0 as

$$\hat{z} = \lambda^T z,$$

where $\lambda = [\lambda_1, \lambda_2, \ldots, \lambda_n]^T$ is an $n \times 1$ matrix of weights and $z = [z(x_1), z(x_2), \ldots, z(x_n)]^T$ the matrix of n observed values that will be used for interpolation at location x_0. We assume the sum of all weights is 1:

$$\lambda^T J = 1,$$

where J is an $n \times 1$ matrix of 1s.

We start from the error of interpolation given by the difference between the estimated value $\hat{z}$ and the true, but unknown, value z at location x_0:

$$e = \hat{z} - z.$$

Of course, we do not know the true value. But the trick here is that we will use a series of transformations to convert the error to another form that can be calculated. Specifically, we will look at the variance of the error at location x_0:

$$\text{var}(e) = E[(e - E[e])^2] = E[e^2] - (E[e])^2.$$

To minimize the variance of the error, we require the expected value of the error to be zero:

$$E[e] = 0,$$

which gives

$$\text{var}(e) = E[e^2] = E[(\hat{z} - z)^2].$$

Using the matrix form of the weighted sum, we have

$$\text{var}(e) = E\left[(\lambda^T z - z)^2\right],$$

which can be expanded as

$$\text{var}(e) = E[z^2] + E[(\lambda^T z)^2] - 2E[z\lambda^T z].$$

We can show that each of the three terms in the above equation can be further expanded and rewritten, using covariances and the stationarity hypothesis. Specifically, we use C

to denote an $n \times n$ covariance matrix where each element, $c_{ij} = \text{cov}(x_i, x_j)$, is the covariance between a pair of observations at locations x_i and x_j, and $\mathbf{C}_0$ to denote an $n \times 1$ covariance matrix where $c_i = \text{cov}(x_0, x_i)$ is the covariance between x_0 and each of the observations. The error variance can then be rewritten as[2]

$$\text{var}(e) = \text{var}(z) + \boldsymbol{\lambda}^T \mathbf{C} \boldsymbol{\lambda} - 2\boldsymbol{\lambda}^T \mathbf{C}_0,$$

where var(z) is the variance of the variable at location x_0. We do not need to worry about how to calculate var(z) because it will be eliminated in the following derivation.

There is one more thing that needs to be considered in the above derivation: we need to explicitly make sure that the sum of the weights is 1, as this assumption is needed several times. In other words, the above derivation process needs to be constrained by the requirement that the sum of weights is 1. We formulate this constraint in a mathematical optimization:

$$\begin{aligned} \text{minimize} \quad & \text{var}(e) \\ \text{subject to} \quad & 2\mu(\boldsymbol{\lambda}^T \mathbf{J} - 1) = 0. \end{aligned}$$

The only constraint in the formulation is to ensure the sum of weights is 1. There is a good reason for adding μ, the mean value at location x_0, here: in this way we can estimate the mean at the same time as we estimate all the weights. This is what we call *ordinary kriging*.

To solve this optimization problem, we treat the equality constraint as a Lagrangian function and move it to the objective function such that our goal becomes to minimize the following new objective function:

$$q = \text{var}(z) + \boldsymbol{\lambda}^T \mathbf{C} \boldsymbol{\lambda} - 2\boldsymbol{\lambda}^T \mathbf{C}_0 - 2\mu(\boldsymbol{\lambda}^T \mathbf{J} - 1).$$

To minimize the variance, we need to make sure two partial derivatives of the new objective function are zero:

$$\frac{\partial q}{\partial \boldsymbol{\lambda}} = 2\mathbf{C}\boldsymbol{\lambda} - 2\mathbf{C}_0 - 2\boldsymbol{\mu} = 0,$$

$$\frac{\partial q}{\partial \mu} = -2\boldsymbol{\lambda}^T \mathbf{J} + 2 = 0,$$

in which $\boldsymbol{\mu}$ is an $n \times 1$ matrix where each element is μ. These two conditions together define an equation system:

$$\begin{cases} \mathbf{C}\boldsymbol{\lambda} - \boldsymbol{\mu} = \mathbf{C}_0 \\ \boldsymbol{\lambda}^T \mathbf{J} = 1 \end{cases},$$

---

[2]The derivation will need to use the property of the covariance between two variables X and Y: $\text{cov}(X, Y) = E[XY] - E[X]E[Y]$ and $\text{var}(X) = \text{cov}(X, X)$. Here we omit the details of the derivation, which can be found in the many references cited at the end of this chapter.

which is a system of $n+1$ variables, and the first set of equations can be expanded as

$$\begin{pmatrix} c_{11} & c_{1,2} & \cdots & c_{1n} & 1 \\ c_{21} & c_{2,2} & \cdots & c_{2n} & 1 \\ \vdots & & & \vdots & \vdots \\ c_{n1} & c_{n2} & \cdots & c_{nn} & 1 \end{pmatrix} \begin{pmatrix} \lambda_1 \\ \lambda_2 \\ \vdots \\ \lambda_n \\ -\mu \end{pmatrix} = \begin{pmatrix} c_{01} \\ c_{02} \\ \vdots \\ c_{0n} \end{pmatrix}.$$

Since we can model the covariances, we can use the above equation system to actually compute the weights along with the estimated mean. However, we can also convert it into a more commonly used form using semivariograms. We recall the relationship between semivariance and covariance, $\gamma(h) = C(0) - C(h)$, and we obtain

$$c_{ij} = c(0) - \gamma_{ij}$$

and

$$c_{0i} = c(0) - \gamma_{0i}.$$

Thus, we can rewrite the above equations as:

$$\begin{pmatrix} \gamma_{11} & \gamma_{1,2} & \cdots & \gamma_{1n} & 1 \\ \gamma_{21} & \gamma_{2,2} & \cdots & \gamma_{2n} & 1 \\ \vdots & & & \vdots & \vdots \\ \gamma_{n1} & \gamma_{n2} & \cdots & \gamma_{nn} & 1 \end{pmatrix} \begin{pmatrix} \lambda_1 \\ \lambda_2 \\ \vdots \\ \lambda_n \\ \mu \end{pmatrix} = \begin{pmatrix} \gamma_{01} \\ \gamma_{02} \\ \vdots \\ \gamma_{0n} \end{pmatrix},$$

which has the equivalent matrix form

$$\gamma \lambda + \mu = \gamma_0,$$

where γ is an $n \times n$ matrix with element γ_{ij} being the semivariance between locations x_i and x_j, and γ_0 is an $n \times 1$ matrix of the semivariances between location x_0 and the n observed locations.

We can now change the above notation slightly so that the second equation ($\lambda^T J = 1$) can be included:

$$\begin{pmatrix} \gamma_{11} & \gamma_{1,2} & \cdots & \gamma_{1n} & 1 \\ \gamma_{21} & \gamma_{2,2} & \cdots & \gamma_{2n} & 1 \\ \vdots & & & \vdots & \vdots \\ \gamma_{n1} & \gamma_{n2} & \cdots & \gamma_{nn} & 1 \\ 1 & 1 & \cdots & 1 & 0 \end{pmatrix} \begin{pmatrix} \lambda_1 \\ \lambda_2 \\ \vdots \\ \lambda_n \\ \mu \end{pmatrix} = \begin{pmatrix} \gamma_{01} \\ \gamma_{02} \\ \vdots \\ \gamma_{0n} \\ 1 \end{pmatrix}.$$

Based on the above notation, we can denote the first $(n+1) \times (n+1)$ matrix on the left-hand side by K, the second by $[\lambda, \mu]$, and the $(n+1) \times 1$ matrix on the right-hand side by k. We can then write

$$K[\lambda, \mu] = k.$$

To solve this equation system, we can use the Gauss–Jordan elimination method. However, in computer programs, it is often more convenient to directly use the product of the inverse matrix of K and matrix k to get the values in λ:

$$[\lambda, \mu] = K^{-1} k.$$

After the weights are calculated, the estimated error variance is

$$\hat{\sigma}^2 = [\lambda + \mu]^T k.$$

The Python program in Listing 8.11 implements the process of ordinary kriging. The function takes two inputs: the data for the nearby locations, and the semivariogram. Here we once again use convenient features from the NumPy module of Python. Line 17 uses the vectorization feature of the `semivariogram` function (passed to here as parameter `model`) to return the semivariances for each distance from the known point to the point to be estimated. Then we make sure k is a 1 by N matrix with the last element (row) being 1 (lines 18 and 20). To compute matrix K, we first compute an N by N distance matrix (line 22). This matrix is flattened into an array of N by N elements that is used to compute the semivariance for each of the distances (line 23). The result is cast back into an N by N matrix (lines 24 and 25). Then we make sure it becomes an N+1 by N+1 matrix with the last row and column being 1s, except for the lower-right corner which is 0 (lines 26–30). Line 31 solves the equation system to obtain the weights, which are used in line 32 to calculate the estimate. Lines 33–37 compute the error standard deviation and make sure this is positive. Finally, the program returns the estimated value, the estimated error standard deviation, estimated mean, and the weights.

Listing 8.11: Ordinary kriging (okriging.py).

```
import numpy as np
from semivariance import distance

def okriging(Z, model):
    """
    Ordinary kriging.
    Input
      Z: an array of [X, Y, Val, Distance to x0]
      model: the name of fitted semivariance model
    Output
      zhat: estimated value at x0
      sigma: standard error
      mu: estimated mean
      w: weights
    """
    N = len(Z)                      # number of points
    k = model(Z[:,3])               # get [gamma(xi, x0)]
    k = np.matrix(k).T              # k is a 1xN matrix
    k1 = np.matrix(1)
```

```
20        k = np.concatenate((k, k1), axis=0) # add a new row of 1s
21        K = [ [distance(Z[i][0:2],Z[j][0:2])
22               for i in range(N)] for j in range(N)]
23        K = np.array(K)                     # list -> NumPy array
24        K = model(K.ravel())                # [gamma(xi, xj)]
25        K = np.matrix(K.reshape(N, N))      # array -> NxN matrix
26        ones = np.matrix(np.ones(N))        # Nx1 matrix of 1s
27        K = np.concatenate((K, ones.T), axis=1) # add a col of 1s
28        ones = np.matrix(np.ones(N+1))      # (N+1)x1 of 1s
29        ones[0, N] = 0.0                    # last one is 0
30        K = np.concatenate((K, ones), axis=0) # add a new row
31        w = np.linalg.solve(K, k)           # solve: K w = k
32        zhat = (np.matrix(Z[:,2])*w[:-1])[0, 0] # est value
33        sigmasq = (w.T * k)[0, 0]           # est error var
34        if sigmasq < 0:
35            sigma = 0
36        else:
37            sigma = np.sqrt(sigmasq)        # error
38        return zhat, sigma, w[-1][0], w     # est, err, mu, w
```

Then we test two points for interpolation (Listing 8.12). First, we use an arbitrary point at location (337000, 4440911); the estimated value is 860.314 with a standard error of 6.757. We then test the first point in the data (note the data point is not taken off the data); the estimated value is 869.0 with 0 error standard deviation. It is obvious that the ordinary kriging method will return exactly the same value if the location has an observation. This feature, called exact interpolation, can be proved formally by working on the equations presented above.

Listing 8.12: Testing ordinary kriging (ordinary_kriging_test.py).

```
1  import numpy as np
2  import sys
3  sys.path.append('../geom')
4  from point import *
5  from fitsemivariance import *
6  from semivariance import *
7  from covariance import *
8  from read_data import *
9  from prepare_interpolation_data import *
10 from okriging import *
11
12 Z = read_data('../data/necoldem.dat')
13
14 hh = 50
15 lags = range(0, 3000, hh)
16 gamma = semivar(Z, lags, hh)
17 covariance = covar(Z, lags, hh)
18
```

```
19  Z1 = np.array(Z)
20  semivariogram = fitsemivariogram(Z1, gamma, spherical)
21  #semivariograml = fitsemivariogram(Z1, gamma, linear)
22  #semivariogramg = fitsemivariogram(Z1, gamma, gaussian)
23
24  if __name__ == "__main__":
25      x = Point(337000, 4440911)
26      P1 = prepare_interpolation_data(x, Z1)[0]
27      print okriging(np.array(P1), semivariogram)[0:2]
28
29      x = Point(Z1[0,0], Z1[0,1])
30      P1 = prepare_interpolation_data(x, Z1)[0]
31      print okriging(np.array(P1), semivariogram)[0:2]
```

8.2.4 Simple kriging

In simple kriging, we take the mean to be constant all over the area. This is different from ordinary kriging where the mean varies from point to point. In this case, we are more interested in what is left if the mean is subtracted from the value: since the mean is always the same, having it in the data does not help much. For this reason, we define a residual function at each location x as follows:

$$R(x) = z(x) - \mu,$$

where μ is the constant mean. Based on this we can define $\boldsymbol{R}$ as an $n \times 1$ matrix where $R_i = R(x_i)$ is the residual at location x_i. We also use $R = R(x_0)$ and $\hat{R} = \hat{R}(x_0)$ to simplify our notation.

Similar to what we did in ordinary kriging, we look at the estimation error:

$$\begin{aligned} e &= \hat{z} - z \\ &= (\hat{z} - \mu) - (z - \mu) \\ &= \hat{R} - R \\ &= \lambda^T \boldsymbol{R} - R. \end{aligned}$$

We can write the error variance as

$$\begin{aligned} \mathrm{var}(e) &= E[(\lambda^T \boldsymbol{R} - R)^2] \\ &= E[R^2] + E[(\lambda^T \boldsymbol{R})^2] - 2E[R\lambda^T \boldsymbol{R}] \\ &= \mathrm{var}(R) + \lambda^T \boldsymbol{C} \lambda - 2\lambda^T \boldsymbol{C}_0. \end{aligned}$$

Since we have already used the mean as a constant to derive the above equation, we will directly take the partial derivatives of the error variance with respect to each of the weights; the variance is minimized if all the derivatives are zero. This is what we call *simple kriging*. We have

$$\frac{\partial \text{var}(e)}{\partial \lambda} = 2C\lambda - 2C_0 = 0.$$

Using the relationship between semivariance and covariance, we can rewrite the above equation system as

$$\gamma \lambda = \gamma_0,$$

which can be written in matrix form as

$$\begin{pmatrix} \gamma_{11} & \gamma_{1,2} & \cdots & \gamma_{1n} \\ \gamma_{21} & \gamma_{2,2} & \cdots & \gamma_{2n} \\ \vdots & & & \vdots \\ \gamma_{n1} & \gamma_{n2} & \cdots & \gamma_{nn} \end{pmatrix} \begin{pmatrix} \lambda_1 \\ \lambda_2 \\ \vdots \\ \lambda_n \end{pmatrix} = \begin{pmatrix} \gamma_{01} \\ \gamma_{02} \\ \vdots \\ \gamma_{0n} \end{pmatrix}.$$

To be consistent with the ordinary kriging formula, we rewrite this as

$$K\lambda = k,$$

where k and K are $n \times 1$ and $n \times n$ matrices, respectively. We use the product of the inverse matrix of K and matrix k to solve the equations and to get the λ values:

$$\lambda = K^{-1}k.$$

After the weights are calculated, the estimated variance of error is

$$\hat{\sigma}^2 = \lambda^T k.$$

The Python code in Listing 8.13 is designed to implement the simple kriging procedure. It is important to note here that to interpolate the unknown value, we need to provide the global mean. Also, we do not directly use the raw data to compute the estimate. Instead, we use the residuals of the data (line 25) to estimate the residual of the unknown location. The estimated value is the sum of the residual and the mean (line 27).

Listing 8.13: Simple kriging (skriging.py).

```
import numpy as np
from semivariance import distance

def skriging(Z, mu, model):
    """
    Simple kriging
    Input
        Z: an array of [X, Y, Val, Distance to x0]
        mu: mean of the data
        model: the name of fitted semivariance model
```

```
11        Output
12          zhat: the estimated value at the target location
13          sigma: standard error
14          w: weights
15        """
16        N = len(Z)                            # number of points
17        k = model(Z[:,3])                     # get [gamma(xi, x0)]
18        k = np.matrix(k).T                    # 1xN matrix
19        K = [ [distance(Z[i][0:2],Z[j][0:2])
20                for i in range(N)] for j in range(N)]
21        K = np.array(K)                       # list -> NumPy array
22        K = model(K.ravel())                  # [gamma(xi, xj)]
23        K = np.matrix(K.reshape(N, N))        # array -> NxN matrix
24        w = np.linalg.solve(K, k)             # solve K w = k
25        R = Z[:,2] - mu                       # get residuals
26        zhat = (np.matrix(R)*w)[0, 0]         # est residual
27        zhat = zhat + mu                      # est value
28        sigmasq = (w.T*k)[0, 0]               # est error variance
29        if sigmasq<0:
30            sigma = 0
31        else:
32            sigma = np.sqrt(sigmasq)          # error
33        return zhat, sigma, w                 # est, error, weights
```

We can once again write a Python program to wrap the `skriging` function with the data (Listing 8.14), where we reuse the data preparation part in the testing of ordinary kriging.

Listing 8.14: Testing simple kriging using the sample data (simple_kriging_test.py).

```
1  from ordinary_kriging_test import *
2  from skriging import *
3
4  if __name__ == "__main__":
5      x = Point(337000, 4440911)
6      P1, mu = prepare_interpolation_data(x, Z1)
7      print skriging(np.array(P1), mu, semivariogram)[0:2]
8
9      x = Point(Z1[0,0], Z1[0,1])
10     P1, mu = prepare_interpolation_data(x, Z1)
11     print skriging(np.array(P1), mu, semivariogram)[0:2]
```

The results of testing simple kriging are shown below. The estimation at the first location is 860.344 with an error of 6.758, which is slightly higher than the previous estimate. The code reports the same results as in the data when the first point in the data is used. For the second point, the method still shows the exact value as in the data.

```
(860.34415501300725, 6.7576953251206504)
(869.0, 0)
```

8.3 Using interpolation methods

We will now use the inverse distance weighted method and kriging to interpolate the surface from the sample data. The program in Listing 8.15 uses the code we have discussed so far (lines 1–19) and shows how to interpolate the surface using the interpolation methods we have discussed so far. Here, we want to create surfaces with a cell size that is 1/100 of the smaller dimension of the area (line 24). In our case, the resolution comes to 30 meters, which gives surfaces 100 cells wide and 100 cells high. We use two-dimensional arrays to hold the results from interpolation (lines 27–31), where each of the kriging methods has both a value surface and an error surface. The process of generating the surfaces uses the location of each point (line 34) to prepare the data (line 35) used in each of the interpolation methods (lines 37–43). We save the data to corresponding files and use an open source tool called gnuplot to draw the surfaces. We create an additional file that stores the surface of differences between the two kriging methods.

Listing 8.15: Interpolating surface (interpolate_surface.py).

```
1  import sys
2  sys.path.append('../geom')
3  from point import *
4  import numpy as np
5  from semivariance import *
6  from fitsemivariance import *
7  from prepare_interpolation_data import *
8  from read_data import *
9  from skriging import *
10 from okriging import *
11 from idw import *
12
13 Z = read_data('../data/necoldem.dat')
14 Z = np.array(Z)
15
16 hh = 50
17 lags = range(0, 3000, hh)
18 gamma = semivar(Z, lags, hh)
19 semivariogram = fitsemivariogram(Z, gamma, spherical)
20
21 x0, x1 = Z[:,0].min(), Z[:,0].max()
22 y0, y1 = Z[:,1].min(), Z[:,1].max()
23 dx, dy = x1-x0, y1-y0
24 dsize = min(dx/100.0, dy/100.0)
25 nx = int(np.ceil(dx/dsize))
26 ny = int(np.ceil(dy/dsize))
27 surfaceOK = np.zeros((nx, ny))
28 errorOK = np.zeros((nx, ny))
29 surfaceSK = np.zeros((nx, ny))
30 errorSK = np.zeros((nx, ny))
31 surfaceIDW = np.zeros((nx, ny))
```

```
32   for i in range(nx):
33   for j in range(ny):
34       x = Point(x0+i*dsize, y0+j*dsize)
35       Z1, mu = prepare_interpolation_data(x, Z)
36       Z1 = np.array(Z1)
37       kresult = okriging(Z1, semivariogram)
38       surfaceOK[i,j] = kresult[0]
39       errorOK[i,j] = kresult[1]
40       kresult = skriging(Z1, mu, semivariogram)
41       surfaceSK[i,j] = kresult[0]
42       errorSK[i,j] = kresult[1]
43       surfaceIDW[i,j] = IDW(Z1, 1.5)
44
45   f1 = open('results/surfaceOK', 'w')
46   f2 = open('results/errorOK', 'w')
47   f3 = open('results/surfaceSK', 'w')
48   f4 = open('results/errorSK', 'w')
49   f5 = open('results/surfaceIDW', 'w')
50   f6 = open('results/surfaceKDiff', 'w')
51   for j in range(ny):
52       for i in range(nx):
53           f1.write(str(surfaceOK[i, j])+' ')
54           f2.write(str(errorOK[i, j])+' ')
55           f3.write(str(surfaceSK[i, j])+' ')
56           f4.write(str(errorSK[i, j])+' ')
57           f5.write(str(surfaceIDW[i, j])+' ')
58           f6.write(str(surfaceOK[i, j]-surfaceSK[i, j])+' ')
59       f1.write('\n')
60       f2.write('\n')
61       f3.write('\n')
62       f4.write('\n')
63       f5.write('\n')
64       f6.write('\n')
65
66   f1.close()
67   f2.close()
68   f3.close()
69   f4.close()
70   f5.close()
71   f6.close()
```

The value surfaces generated by ordinary and simple kriging are shown in Figure 8.9. The first thing to notice is the close similarity between these two sets of results. What difference there is is shown in Figure 8.10. The range of the values in our data is from 12.1491 to 16.9583, and the figure shows the difference between ordinary and simple kriging to be roughly from −0.1 to 0.25, which is in the order of 2 percent. It is also noticeable in Figure 8.10 that there are clusters in this difference map.

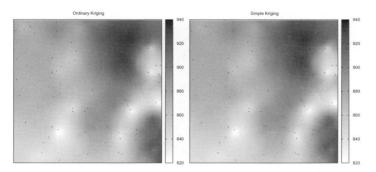

Figure 8.9 Surface generated using ordinary (left) and simple (right) kriging method

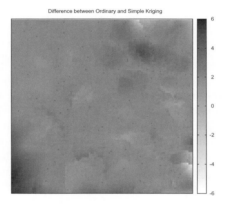

Figure 8.10 Difference between results from ordinary and simple kriging

Figure 8.11 shows the corresponding error surfaces of the two kriging methods. These two maps again show very similar patterns. An obvious trend is the coincidence between the location of low errors and the observed data point. This figure demonstrates why kriging is called an exact interpolation method: it returns the observed value at locations with known values.

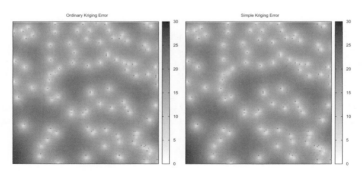

Figure 8.11 Error of estimation

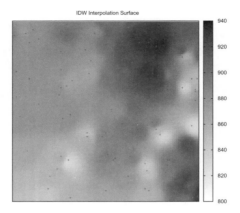

Figure 8.12 Surface generated using the inverse distance weighted method

Finally, the interpolated surface generated by IDW is shown in Figure 8.12. This map is noticeably different from those generated by the kriging methods. A careful examination of the IDW map also shows the influence of the observation data on the results: there is a strong correlation between each observation and its nearby interpolation result, shown as the bull's-eyes (the light or dark circular areas) around those points. While the same effect can also be found on the kriging surfaces, on the IDW surface it is more obvious. In other words, kriging surfaces tend to be smoother than IDW surfaces.

So, which method, kriging or IDW, provides the better result? The theoretical error surface returned by kriging does not really tell us the quality of this method, especially when compared with other methods such as IDW. Here, we use the sample cross-validation approach as we did for the IDW to see how much error the kriging method creates on those locations where we know the result (but the known value is not used as the input to create the semivariance or kriging). The code in Listing 8.16 implements this and returns an RMSE value of 13.2574012443. What would happen if we used the same semivariogram calculated using the entire data set but mask each point for interpolation? We can make some slight changes to the code so that we do not calculate gamma and the semivariogram each time. In this case, the RMSE is 13.4853542512. It should also be noted that we only use the nearest neighbors to compute the semivariance (i.e., we use P1 instead of P in line 27). This of course will affect (increase) the overall RMSE, but we can gain performance by decreasing the running time. In any of these settings, we obtained an RMSE that is close to the best of IDW (13.2774718585).

Listing 8.16: Cross-validation for ordinary kriging (kriging_cross_validation.py).

```
1  import numpy as np
2  import sys
3  sys.path.append('../geom')
4  from point import *
5  from fitsemivariance import *
6  from semivariance import *
7  from covariance import *
```

Interpolation

```
8   from read_data import *
9   from prepare_interpolation_data import *
10  from okriging import *
11
12  Z = read_data('../data/necoldem250.dat')
13
14  hh = 50
15  lags = np.arange(0, 3000, hh)
16  test_results = []
17  N = len(Z)
18  mask = [True for i in range(N)]
19  numNeighbors = 10
20
21  for i in range(N):
22      mask[i] = False
23      x = Point(Z[i][0], Z[i][1])
24      P = [ Z[j] for j in range(N) if mask[j] == True]
25      P1 = prepare_interpolation_data(x, P, numNeighbors)[0]
26      P1 = np.array(P1)
27      gamma = semivar(P1, lags, hh)
28      if len(gamma) == 0:
29          Continue
30      semivariogram = fitsemivariogram(P1, gamma, spherical)
31      kresult = okriging(P1, semivariogram)
32      test_results.append(kresult[0]-Z[i][2])
33      mask[i] = True
34
35  print np.sqrt(sum([r**2 for r in test_results])/
36               len(test_results))
```

The above results leave us wondering: is kriging actually a better method in giving more accurate estimates? Of course, we cannot come to a definitive conclusion about this on the basis of one simple example. But what can such a simple example tell us? Let us look at where the data comes from in Figure 8.13. Our 100 points were randomly sampled

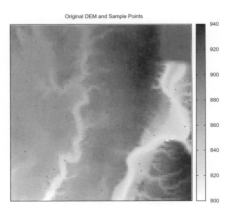

Figure 8.13 The original DEM where the 100 sample points are obtained

from a USGS 7.5 minute Digital Elevation Model (DEM) quadrangle with a resolution of 30 meters. Comparing the original DEM with the one we interpolated above, one thing we can say for sure is they are different, albeit that the interpolated surfaces resemble the original one. What would happen if we increased the sample size?

Increasing the number of points sampled to 250 and running the IDW cross-validation, we get the following result:

```
12.2694123739  [ 0   ]
11.2761773623  [ 0.5 ]
10.4449626353  [ 1   ]
9.88226564121  [ 1.5 ]
9.55568320108  [ 2   ]
9.3882134655   [ 2.5 ]
9.32030718212  [ 3   ]
9.31308533549  [ 3.5 ]
9.34163738372  [ 4   ]
9.38982234683  [ 4.5 ]
9.44722571849  [ 5   ]
```

Sampling a larger number of points definitely increases the overall estimation accuracy. The ordinary kriging method gives us a cross-validation RMSE of 8.66773948442. This time, kriging is clearly better. But we will need more experiments like this to gain more insight into the pros and cons of the two methods. Another issue is that the optimal power value in IDW changes to 3.5 as compared to 1.5 in the 100-point sample; kriging does not have such subjectivity because all parameters are estimated using the data. Figure 8.14 shows the new result, which captures detail that cannot be obtained using 100 points.

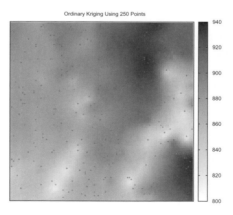

Figure 8.14 The DEM created by ordinary kriging using 250 sample points

8.4 Midpoint displacement

In this section, we extend our discussion of interpolation to an interesting area: how can we generate an artificial and random surface that mimics the real earth terrain? Here, we think of the variation between the elevation values of two locations as a function of the distance between the two locations. More specifically, we use d to denote the distance between two locations x and $x+d$, and we say that the variance of the values between these two locations is

$$E[z(x) - z(x+d)]^2 \propto |d|^{2H},$$

where $0 \leq H \leq 1$ is a constant that controls the overall variation of the values.

Fournier et al. (1982) developed an algorithm called midpoint displacement to recursively add a midpoint to a given set of four points; the values at these four points are samples from a normal distribution with variance σ^2, and we assume the four points are located at the four corners of a square with side length s. The center point is at a distance of $s/\sqrt{2}$ from each of the four corners. The value of the center point of the square is calculated by drawing a sample from a normal distribution with variance $(\frac{1}{2})^H \sigma^2$ and mean the average of the values of the four points. In other words, we add a small displacement that is drawn from a normal distribution $N(0, (\frac{1}{2})^H \sigma^2)$ to the mean of the four values; hence the name of the algorithm. Increasing the H value from 0 to 1 will decrease the variance of the displacement from σ^2 to $\frac{1}{2}\sigma^2$.

Figure 8.15 Midpoint displacement on square (left) and diamond (right) configurations. Black dots represent the set of current points and the light grey dot represents the generated point

There are two spatial configurations for the four points: square and diamond (Figure 8.15). The algorithm starts with the square configuration for the first iteration and alternately uses the two configurations to interpolate additional points to fill the space. At each iteration n, we first conduct interpolations using the square configuration and then the diamond. The number of points on each side at iteration n is $2^n + 1$ and the total number of points is $(2^n + 1)^2$. The variance at iteration n is $(\frac{1}{2})^{nH}\sigma^2$. Figure 8.16 illustrates a process of three iterations that yields a grid of 81 points. The algorithm also adds an additional small value, drawn from the same normal distribution, to the generated values to create more randomness in the result.

The Python code in Listing 8.17 is mainly adopted from pseudo-code published in the book by Peitgen and Saupe (1988). The function `f` (line 7) is used to compute the displacement of the mean value of a list of points using a standard deviation `delta`. The input of point values can be technically of any length, but we will use either a length of 4 as in a typical case where four points are available, or only 3 points for interpolating points at the edge of the grid (e.g., the middle-left point in the upper-right diagram of Figure 8.16).

The function `midpoint2d` conducts the main process of the algorithm. The input parameter `maxlevel` determines the total number of iterations of the algorithm and

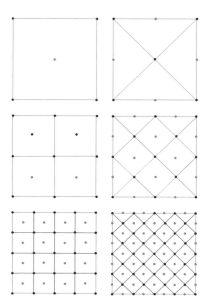

Figure 8.16 Three iterations of midpoint displacement, creating a grid of 81 points

therefore the final size of the grid to be created (line 11). The parameter `sigma` is the initial standard deviation of the normal distribution of the values. Lines 14–17 initialize the four corners of the grid, all drawn from a normal distribution with mean 0 and standard deviation specified in the `midpoint2d` input. In each iteration, we use variables `D` and `d` to decide which points to interpolate, and the initial values of these variables are set in lines 18 and 19. At each iteration of the process, the standard deviation is recalculated for the square (line 21) and diamond (line 30) configurations. For the square configuration, we always have four points to interpolate the interior, as done within the `for` loop from line 23. Additional randomness can be added from line 26. For the diamond configuration, however, we need to compute the points at the boundary with three input points. This is done within the `for` loop from line 31. The interior points under the diamond configuration where four points are available are computed in the blocks beginning with lines 36 and 40. Then additional randomness can be added (line 44).

Listing 8.17: Fractal interpolation (midpoint2d.py).

```
1  import random
2  import math
3  import numpy as np
4  import sys
5  import string
6
7  def f(delta, x):
8      return sum(x)/len(x)+delta*random.gauss(0,1)
9
```

Interpolation

```python
def midpoint2d(maxlevel, sigma, H, addition=False):
    N = int(math.pow(2, maxlevel))
    X = [ [0]*(N+1) for x in range(N+1)]
    delta = sigma
    X[0][0] = delta*random.gauss(0,1)
    X[0][N] = delta*random.gauss(0,1)
    X[N][0] = delta*random.gauss(0,1)
    X[N][N] = delta*random.gauss(0,1)
    D = N
    d = N/2
    for stage in range(1, maxlevel+1):
        delta = delta*math.pow(0.5, 0.5*H)
        for x in range(d, N-d+1, D):
            for y in range(d, N-d+1, D):
                X[x][y] = f(delta, [X[x+d][y+d], X[x+d][y-d],
                                    X[x-d][y+d], X[x-d][y-d]])
        if addition is True:
            for x in range(0, N+1, D):
                for y in range(0, N+1, D):
                    X[x][y] += delta*random.gauss(0, 1)
        delta = delta*math.pow(0.5, 0.5*H)
        for x in range(d, N-d+1, D):
            X[x][0] = f(delta,[X[x+d][0],X[x-d][0],X[x][d]])
            X[x][N] = f(delta,[X[x+d][N],X[x-d][N],X[x][N-d]])
            X[0][x] = f(delta,[X[0][x+d],X[0][x-d],X[d][x]])
            X[N][x] = f(delta,[X[N][x+d],X[N][x-d],X[N-d][x]])
        for x in range(d, N-d+1, D):
            for y in range(D, N-d+1, D):
                X[x][y] = f(delta, [X[x][y+d], X[x][y-d],
                                    X[x+d][y], X[x+d][y]])
        for x in range(D, N-d+1, D):
            for y in range(d, N-d+1, D):
                X[x][y] = f(delta, [X[x][y+d], X[x+d][y-d],
                                    X[x+d][y], X[x-d][y]])
        if addition is True:
            for x in range(0, N+1, D):
                for y in range(0, N+1, D):
                    X[x][y] += delta*randomgauss(0, 1)
            for x in range(d, N-d+1, D):
                for y in range(d, N-d+1, D):
                    X[x][y] += delta*random.gauss(0, 1)
        D=D/2
        d=d/2
    return X

if __name__ == '__main__':
    """
    Python midpoint2d.py
    Python midpoint2d.py maxlevel sigma H
```

```
59          """
60          maxlevel = 6
61          sigma = 0.5
62          H = 0.1
63          if len(sys.argv) == 4:
64              maxlevel = string.atoi(sys.argv[1])
65              sigma = string.atof(sys.argv[2])
66              H = string.atof(sys.argv[3])
67          X = midpoint2d(maxlevel, sigma, H)
68          for i in X:
69              for j in i:
70                  print j,
71              print
```

Figure 8.17 shows three surfaces generated using the algorithm. All three cases have the same initial standard deviation of 0.5. It is clear that a high H value leads to a smooth surface (left). The surface tends to become rougher as the H value decreases.

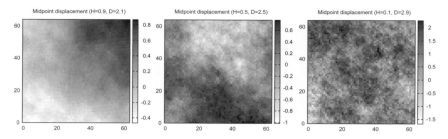

Figure 8.17 Surfaces created using the midpoint displacement algorithm with different H values

8.5 Notes

What exactly is kriging? The kriging method is also called optimal prediction or interpolation, due to its minimization of error variance. We have shown how kriging is used to estimate unknown values based on a given set of observations. The discipline called geostatistics started in the 1950s from the empirical work of Danie G. Krige, a South African mining engineer, who developed a probabilistic approach to estimating the density of mineral resources (Krige, 1951). However, Krige's work received little attention until the French engineer Matheron (1963) gave the name "kriging" to this spatial interpolation method in honor of Krige's original work. In the account by Cressie (1990), different scholars have defined kriging in different ways, but most definitions recognize the weighting feature of this method. Since the 1980s, geostatistics has become a well-defined discipline in geoscience. There have been many books dedicated to this topic in different application domains such as geology and natural resources (Webster and Oliver, 1990; Cressie, 1991; Carr, 1995); readers requiring more detail can refer to these.

Besides the simple and ordinary kriging we have focused on in this book, there are other kinds of kriging methods that have been developed to deal with different situations. For example, universal kriging can handle a general trend in the data. In our description above, the semivariogram does not change when the direction to the known points changes. We can certainly accommodate spatial anisotropy (meaning that means are different in different directions) by allowing such changes. It is also possible to take into account the spatial relationship between the variable of interest and other spatial variables which may provide more information in making better predictions, a premise of cokriging.

Beyond kriging and inverse distance weighting, there are other methods that can be used in interpolation. The first law of geography (Tobler, 1970) has always been a theoretical underpinning of these methods. Burrough and McDonnell (1998) gave a list of interpolation methods and compared their strengths and weaknesses. According to them, inverse distance weighting can be used for a "quick interpolation from sparse data on [a] regular grid or irregularly spaced samples" with "small" computing load. They argued that, with a "moderate" computing load, "when data are sufficient to compute variograms, kriging provides a good interpolator for sparse data. Binary and nominal data can be interpolated with indicator kriging. Soft information can also be incorporated as trends or stratification. Multivariate data can be interpolated with cokriging." Later in this book we will visit some other methods that can also be used for interpolation. One of these methods is Voronoi polygons, where we assume the entire area that is nearest to an observation point has the same value as the observation.

Another issue that is worth discussing is the type of data for interpolation. We have been dealing with point data where the observations are made at point locations and we want to compute the value at other point locations. What happens when we are dealing with lines? For example, interpolation methods have a long history in cartography in creating contour lines (Xiao and Murray, 2015) as a way to represent the terrain.

We also need to mention time: are the interpolation methods time-consuming? When we look at the outer loop of the algorithms, we need to go through each of the given points once, so that we are actually talking about a linear time complexity. However, when it comes to data preparation, there is a heavy workload: we need to find the N nearest points around the location to be estimated. In this chapter, we have used a brute-force approach. We will see how spatial indexing methods can help greatly improve the performance in later chapters.

Finally, we observe that the midpoint displacement algorithm generates a surface where points that are near to each other tend to have a similar amount of variation that is determined by the standard deviation of the normal distribution used. A surface created by this approach exhibits a feature called self-similarity, where if we enlarge a small portion of the surface generated we can observe the same traits as found in the original whole area. There is a huge body of literature devoted to this type of phenomenon, called fractals (Mandelbrot, 1967, 1977a,b; Peitgen et al., 1992). The metric that is used to measure such self-similarity is called the fractal dimension, which in our method can be simply calculated as $D = 3 - H$. These are the numbers shown in each of the surfaces in Figure 8.17.

8.6 Exercises

1. In the kriging code provided in this chapter, the power model is not included. Write the Python code for this model and compare it with the other models using our example data.

2. We can try a reverse-engineering approach to examine the artificial surface generated using the midpoint displacement algorithm. The basic idea is to first create a surface using the algorithm. We then hold a few points out from the generated surface and use the kriging method or IDW to interpolate those points. How close will the interpolation results be? In kriging, what will be the best model to fit the data?

3. The two kriging methods (simple and ordinary) introduced in this chapter are not applicable when there is a trend in the data. The trend can be fitted using a polynomial function, or other functions such as cubic or even just linear. Subtracting the trend value from the original value will return the error that can then be modeled using a kriging method. This is called universal kriging. Read more about universal kriging (Cressie, 1991) and write a Python program to implement it.

9

Spatial Pattern and Analysis

Grown-ups love figures. When you tell them that you have made a new friend, they never ask you any questions about essential matters. They never say to you, "What does his voice sound like? What games does he love best? Does he collect butterflies?" Instead, they demand: "How old is he? How many brothers has he? How much does he weigh? How much money does his father make?" Only from these figures do they think they have learned anything about him.

If you were to say to the grown-ups: "I saw a beautiful house made of rosy brick, with geraniums in the windows and doves on the roof," they would not be able to get an idea of that house at all. You would have to say to them: "I saw a house that cost $20,000." Then they would exclaim: "Oh, what a pretty house that is!"

Antoine de Saint-Exupéry, *The Little Prince*

There are things we can say about the distribution of spatial data. Ironically, as in the story told in the lovely book *The Little Prince*, we will have to behave like grown-ups who are more interested in numbers than in other kinds of descriptions about things. We would typically consider a number or a set of numbers to describe what we can observe from the spatial data, and we call this spatial analysis. The scope of spatial analysis is broad, but we will focus on a very particular task of spatial analysis: spatial patterns. Figure 9.1 highlights a subset of the trees in the Oval on the Ohio State campus. Are those trees randomly distributed? Before we answer such a question, though, it will be important to ask why it is so important to know whether those trees are randomly distributed. One may be curious about the general landscaping principle in this case. However, there are more important aspects of questions like this. Suppose we have a set of crime locations; whether the crime scenes are random or not will be significant. If the points are not random, they could be clustered so that there is a more of a tendency to see more points around certain locations compared to others. All of these scenarios have implications that may require further investigation to understand the cause and process of the events.

There are many methods and algorithms designed for pattern analysis on various data types. We first introduce a couple of methods that analyze point patterns. Then we move on to the patterns of areal phenomena. An important use of pattern analysis is in landscape ecology, and we discuss one of the metrics in that area. We then extend our discussion of patterns by introducing a clustering method that can be used to help us identify clusters in spatial data.

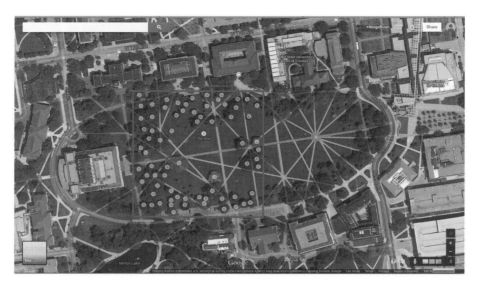

Figure 9.1 Trees in the Oval on the Ohio State campus. Trees are identified using a visual examination of the canopy on the image

9.1 Point pattern analysis

Sometimes we only care about the occurrences of events, and not the values associated with those events. For example, locations of crimes alone may be of interest, but not the type of crime or the damage caused by the crime. Or we might simply be interested in tree locations, and not in their heights. In this sense, we are talking about point pattern analysis. Various methods have been developed for this purpose. We introduce two of them in this section.

9.1.1 Nearest neighbor analysis

Let $P = \{p_1, p_2, \ldots, p_n\}$ be a set of points and d_i be the distance between the ith point and its nearest neighbor. We calculate the average nearest distance for all points as $R_0 = \frac{1}{n}\sum_{i}^{n} d_i$. What would we expect R_0 to be? We compare R_0 with a random case where all the points are randomly distributed in the area. What would the average distance between nearest neighbors be if the points are randomly distributed? That, of course, depends on the number of points or the density of points in the area. Let us assume a case where the point events in an area exhibit a complete spatial randomness, meaning the location of each point is independent of other points and follows a Poisson distribution. Given any location, the probability of no points falling within a radius of x is $e^{-\lambda \pi x^2}$, where λ is the density of points in the area. Therefore the probability of having a nearest neighbor distance X that is smaller than the radius x is $\Pr(X \leq x) = 1 - e^{-\lambda \pi x^2}$. This leads to an exponential distribution of πX^2, and it can be derived that the expected distance of X is $1/(2\sqrt{\lambda})$ with a variance of $(4-\pi)/4\lambda\pi$.

Suppose there are n points in an area of size A. The density of points is $\lambda = n/A$ and the expected average nearest neighbor distance therefore is $R_e = \dfrac{1}{2}\sqrt{A/n}$, and we can calculate a nearest neighbor statistic

$$R = \frac{R_0}{R_e}.$$

An R value that is significantly greater than 1 means the points are dispersed and a value smaller than 1 indicates a clustered pattern. Otherwise, if the R value is around 1, we are expecting a random pattern. But how much is significant? Since we know the mean and variance of the expected value, we can standardize the R value as:

$$z = \frac{R-1}{\sqrt{(4-\pi)/4\lambda\pi}}.$$

A t-test can be used to help us evaluate the pattern. For a confidence level of 0.95, a z value smaller than -1.96 indicates a clustered pattern, a value greater than 1.96 means a dispersed pattern, and a random pattern should have a value in between.

The Python code in Listing 9.1 calculates the nearest neighbor distance statistic. In the function nnd (line 12), we use the k-D tree we developed before to index the points (line 12) and the nearest neighbor search algorithm to find the nearest neighbor of each point (line 25). With the indexing and search algorithm implemented already in Section 5.1, the code for the nearest neighbor distance statistic is short and straightforward, returning the values of mean and expected nearest distances (R_0 and R_e, respectively) and their ratio (R), along with the variance and the standardized z score.

Listing 9.1: Nearest neighbor distance (nnd.py).

```
1  from math import sqrt, pi
2  import numpy as np
3  import random
4  import sys
5  sys.path.append('../geom')
6  sys.path.append('../indexing')
7  from point import *
8  from extent import *
9  from kdtree1 import *
10 from kdtree3 import *
11
12 def nnd(points, density):
13     """
14     Nearest neighbor distance.
15     Input
16       points: a list of Point objects
17       density: density of the points
18     Return
19       R0, Re, R, Var, z
20     """
21     tree = kdtree2(points)
```

```
22          R0 = 0.0
23          for p in points:
24              neighbor = kdtree_nearest_neighbor_query(tree,
25                                                       p, 2)[1][0]
26              R0 += p.distance(neighbor)
27          R0 /= len(points)
28          Re = 1.0/(2.0*sqrt(density))
29          R = R0/Re
30          Var = (4-pi)/(4*density*pi)
31          z = (R-1)/sqrt(Var)
32          return R0, Re, R, Var, z
33
34    def test():
35        n = 100
36        xmin = 10
37        ymin = 10
38        xmax = 40
39        ymax = 40
40        area = Extent(xmin=xmin, xmax=xmax, ymin=ymin, ymax=ymax)
41        density = float(n)/area.area()
42
43        # regular
44        nrow = 10
45        ncol = 10
46        dhorizontal = float((ymax-ymin))/ncol
47        dvertical = float((xmax-xmin))/nrow
48        xs = [ ymin + dvertical/2.0 + dvertical*i
49               for i in range(ncol) ]
50        ys = [ xmin + dvertical/2.0 + dvertical*i
51               for i in range(nrow) ]
52        points = [ Point(x, y) for x in xs for y in ys]
53        R0, Re, R, Var, z = nnd(points, density)
54        print "Uniform:", R0, Re, R, z
55
56        # random
57        points = [ Point(random.uniform(xmin, xmax),
58                         random.uniform(ymin, ymax))
59               for i in range(n) ]
60        R0, Re, R, Var, z = nnd(points, density)
61        print "Random:", R0, Re, R, z
62
63        # clustered
64        points = [ Point(random.gauss(25, 2),
65                         random.gauss(25, 2))
66               for i in range(n) ]
67        R0, Re, R, Var, z = nnd(points, density)
68        print "Gaussian:", R0, Re, R, z
69
70    if __name__ == "__main__":
71        test()
```

In the `test` function, we use the `nnd` function on three sets of data: a regular case where the points are evenly distributed in the space with a fixed interval (line 52); a random case where the X and Y coordinates of each point follow a uniform random distribution (line 59); and a clustered case where the points are around the center of the area with each coordinate following a Gaussian distribution (line 66). In the first case, the points are evenly placed in the area and we can regard this as a case of a dispersed pattern where, metaphorically speaking, the points try to avoid each other. Running the Python program yields the following results, where the four values in each line are R_0, R_e, R, and z, respectively.

```
$ python nnd.py
Uniform: 3.0869848481 1.5 2.05798989873 1.34933096134
Random: 1.81649019531 1.5 1.21099346354 0.269095209073
Gaussian: 0.438146681302 1.5 0.292097787535 -0.90283883998
```

The test results seem to make sense, as the uniform (dispersed) case returns a large nearest neighbor distance while the clustered case has a very small distance, compared with the random case. But the z scores are all compared with a theoretical random case, and how far should we trust the theoretical derivation of the average distance? Let us first consider an informal way to make sense of the expected nearest neighbor distance when the points are randomly distributed. Specifically, we want to know what is the average size of the area where we are expected to find a point. For the n points in an area of size A, we would expect to find a point in an area of A/n, which is the inverse of the density, $1/\lambda = A/n$. If this area is a square, the length of each side of the square is therefore $1/\sqrt{\lambda}$. Suppose the point is at the center of the square; where is the nearest location we would expect another point to appear? Right at the middle of any side, at a distance of exactly $1/\sqrt{\lambda}$ from the center of square.

For a randomly selected point, using our informal reasoning, the point sits in the center of a square of size $1/\sqrt{\lambda}$. But the same goes for the other points: they can all sit in the center of their own squares. Would this not make the distance between points $1/\sqrt{\lambda}$ instead of $1/(2\sqrt{\lambda})$? This is sound thinking, but the numbers are not quite correct. We will not pursue the theoretical derivation any further here, partly to avoid an endless debate on the meaning and implication of complete spatial randomness. Instead, we can compute what an average distance would look like. We use a Monte Carlo simulation to run a large number of simulations; in each simulation the points are placed in the area randomly[1] and the R_0 value is calculated empirically. This is implemented in function `nnd_monte_carlo` in Listing 9.2. By default, we run 10,000 simulations, but this can be changed as an input. The function returns the 2.5th, 50th, and 97.5th percentiles which give a good estimate of the median (50th percentile) and confidence interval.

---

[1] In fact pseudo randomly, in the sense that the calculations are done using a random number generator in Python.

Listing 9.2: Monte Carlo test of the nearest neighbor distance (nnd_monte_carlo.py).

```
from nnd import *

def nnd_monte_carlo(n, area, verbal=False, N=10000):
    Rm = []
    density = float(n)/area.area()
    for i in range(N):
        points = [ Point(random.uniform(area.xmin, area.xmax),
                         random.uniform(area.ymin, area.ymax))
                   for i in range(n) ]
        Rm.append(nnd(points, density)[0])
    Rm = np.array(Rm)
    if verbal:
        for r in Rm:
            print r
    return np.percentile(Rm, 2.5),\
        np.percentile(Rm, 50),\
        np.percentile(Rm, 97.5)

if __name__ == "__main__":
    n = 100
    xmin = 10
    ymin = 10
    xmax = 40
    ymax = 40
    area = Extent(xmin=xmin, xmax=xmax, ymin=ymin, ymax=ymax)
    Rm = nnd_monte_carlo(n, area)
    print Rm
```

The result of running the Monte Carlo code is as follows:

(1.5492612728187518, 1.7516694945432549, 1.9620854828335306)

It is clear that the empirical mean nearest distance for 100 points in the area we use is not 1.5 but about 1.75, which is consistent with the logic discussed above. It is also clear that the random points exhibit an average nearest neighbor distance that is well within the 95 percent confidence interval, while both clustered and dispersed patterns are outside the interval. By setting the `verbal` parameter to `True` we can print out all the distances for each of the 10,000 samples. We plot a histogram of these distances (Figure 9.2) that gives a visual test of these relationships.

We now test the nearest neighbor distance statistic using the trees in the Oval (Figure 9.1). We first digitize all the marked tree points using their screen coordinates. These are the 72 pairs of coordinates stored in the list called `rawpoints` in Listing 9.3. It should be noted that the trees are identified approximately, as some canopies are too close to each other and it is difficult to separate all the trees; some very small trees are not included either. The rest of the lines in the listing transform the points from the screen

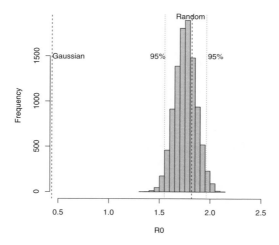

Figure 9.2 The histogram of average nearest neighbor distances for 10,000 simulations for 100 random points with X and Y coordinates ranging from 10 to 40. The two thick dashed lines represent two data sets with 100 points around point (25, 25) from a Gaussian distribution (variance is 2 for both coordinates), and from a uniform distribution in the area

coordinate system (where Y increases from top to bottom) to a conventional Euclidean coordinate system where the origin is at the lower-left corner and the total extent of the points is stretched into a square of 6 by 6 units that includes a padding of 0.1 units on all sides. The last line in the listing converts these coordinates into a list of Point objects that are subsequently used in other calculations.

Listing 9.3: Data points for trees in the Oval (oval_trees.py).

```
1  import sys
2  sys.path.append('../geom')
3  from point import *
4
5  # screen coordinates of the points
6  rawpoints = [
7      [3.99, 4.15], [3.81, 3.33], [5.26, 2.98], [3.64, 4.46],
8      [4.86, 4.50], [3.33, 4.22], [3.01, 4.30], [3.21, 4.46],
9      [3.09, 4.62], [3.40, 4.46], [3.36, 4.93], [3.33, 5.21],
10     [3.52, 5.17], [3.13, 5.21], [5.11, 2.66], [3.80, 5.21],
11     [3.80, 5.09], [3.96, 5.01], [3.80, 4.85], [4.15, 5.05],
12     [4.14, 5.21], [4.61, 5.17], [4.65, 5.05], [4.49, 5.01],
13     [4.81, 4.85], [5.04, 5.05], [5.36, 5.00], [5.38, 5.26],
14     [5.58, 5.18], [5.74, 5.14], [5.78, 4.91], [5.66, 4.16],
15     [5.54, 3.65], [5.38, 2.74], [3.65, 2.74], [4.79, 3.10],
16     [4.44, 3.10], [4.52, 2.98], [4.67, 2.70], [4.36, 2.86],
17     [3.93, 2.94], [4.08, 2.82], [4.16, 2.66], [4.32, 2.51],
18     [3.96, 2.63], [3.69, 2.55], [3.33, 2.66], [3.33, 2.98],
```

```
19          [3.06, 3.02], [3.06, 3.77], [3.18, 3.89], [3.26, 3.53],
20          [3.53, 3.57], [4.67, 2.43], [4.95, 2.55], [5.03, 2.43],
21          [5.89, 3.61], [5.93, 3.81], [5.89, 4.04], [5.93, 4.24],
22          [5.66, 2.53], [5.42, 2.53], [5.89, 2.69], [5.70, 2.96],
23          [5.09, 5.31], [3.86, 5.39], [3.46, 4.72], [3.70, 4.25],
24          [3.14, 3.29], [3.02, 2.43], [3.33, 3.33], [3.06, 2.70] ]
25
26  ymax = max([p[1] for p in rawpoints])
27  ymin = min([p[1] for p in rawpoints])
28  xmin = min([p[0] for p in rawpoints])
29  xmax = max([p[0] for p in rawpoints])
30
31  # add 0.1 to all sides to make it 6x6
32  xratio = 5.8/(xmax-xmin)
33  yratio = 5.8/(ymax-ymin)
34
35  # converting screen coordinates to cartesian [0-6]x[0-6]
36  points = [ Point(xratio*(p[0]-xmin)+0.1,
37                   yratio*(ymax-p[1])+0.1)
38             for p in rawpoints]
```

With the trees mapped, we can use the lines in Listing 9.4 to calculate the nearest neighbor distance statistic for the Oval trees. We also want to test the result empirically using the Monte Carlo method.

Listing 9.4: Testing nearest neighbor distance (test_oval_trees_nnd.py).

```
1   from oval_trees import *
2   from nnd import *
3   from nnd_monte_carlo import *
4
5   area = Extent(xmin=0.0, xmax=6.0, ymin=0.0, ymax=6.0)
6   density = len(points)/area.area()
7   R0, Re, R, Var, z = nnd(points, density)
8   print R0, Re, R, Var, z
9   Rm = nnd_monte_carlo(len(points), area)
10  print Rm
```

Testing the trees returns the following results:

```
0.458288702278 0.353553390593 1.29623619649 0.0341549430919
   1.60291812584
(0.35449964236594012, 0.41160208470080117, 0.4703053746438054)
```

We also plot the results in a histogram (Figure 9.3). The average distance is about 0.4583, which is slightly smaller than the 95th percentile sample from the Monte Carlo simulation. So it is probably a random distribution. Although we are not absolutely confident about the random pattern, we can confidently say this is not a cluster pattern because otherwise the average distance should be on the other (left-hand) side of the histogram. If we can accept a confidence level lower than 95 percent, we may even consider it as a dispersed pattern. This result makes sense from a landscaping perspective: we perhaps prefer the trees to cover a large portion of the area in a public space like the Oval.

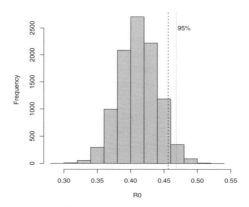

Figure 9.3 Histogram of average nearest neighbor distances for 10,000 simulations for the 72 trees in the Oval

9.1.2 Ripley's *K*-function

Let us think about the design principle of the nearest neighbor distance statistic before we move on to our next topic. The reason we are able to conclude whether a set of points follows a random, clustered, or dispersed pattern is that we can tell how close the points are, or how close they should be for them to be considered as conforming to one of the three patterns. Suppose that, instead of using a fixed way of measuring closeness between points (i.e., using the nearest neighbor), we draw a circle around each point and examine the points enclosed within the circle against what we should expect on a random basis. We can also examine the pattern at different spatial scales by changing the size of the circle. With this comparison, we make our conclusions. This idea leads to the development of Ripley's *K*-function.

Let λ be the density of points in the area of interest. Given a point, we draw a circle of radius d and count the number n of points in the circle (including the given point itself). We define the metric $K(d) = n / \lambda$. What should we expect this metric to be when the points are randomly distributed in the area? The circle has a size of πX^2. If the size of the entire area is A and there are a total of N points in the area, then the circle takes up a ratio of $\pi X^2 / A$ of the entire area. If the points are randomly distributed, we should expect $n = N\pi d^2 / A$ points in the circle and, therefore, $K(d) = (N\pi d^2 / A) / \lambda$. Since $\lambda = N / A$, we have $K(d) = \pi d^2$ for a random pattern. We can technically take out the π and square in this formula by defining the *K*-function as

$$L(d) = \sqrt{\frac{K(d)}{\pi}}.$$

We can then increase the radius of the circle and redo the calculation so that we establish a relationship between $L(d)$ and the radius. For a random pattern, we should expect $L(d) = d$, which is a straight diagonal line when plotting $L(d)$ against d. For a set of many points, we calculate the $L(d)$ value for each point and take the average of all the points for each d value. The fact that we can examine the K-function value against different radii gives us a unique perspective on the distribution of the points at different scales. The $L(d)$ values that are smaller than d indicate a clustered pattern where more points are counted than expected. When the $L(d)$ values are greater than d, we would encounter a smaller number of points than expected and therefore a dispersed pattern. Otherwise, when $L(d)$ equals d, the points exhibit a random pattern.

We can empirically test whether an $L(d)$ value indicates a random pattern using Monte Carlo simulation. The idea here is to generate a large number of random patterns and, for each of the patterns, compute the $L(d)$ values and extract the 2.5th and 97.5th percentiles for each d value, yielding a 95 percent confidence interval.

Computing $L(d)$ and carrying out the Monte Carlo simulations is not difficult, especially with the help of the k-D tree and the circular range search. Listing 9.5 includes functions for both tasks (lines 12 and 31, respectively). Note that the `kfunc` function requires an indexing tree as an input. Unlike the nearest neighbor distance program where the tree is created within the `nnd` function, here we require the tree to be created outside the `kfunc` function. This is because the calculation of $L(d)$ often repeats many times for a set of d values and we only need to create the tree once for a given set of data points. Specifically, the input to the `kfunc` function includes the k-D tree, the point location to be examined, the circle radius, and the density of points in the area (line 12). Once the points in the circle are found (line 25), it is straightforward to obtain the $L(d)$ value. The Monte Carlo simulation function `kfunc_monte_carlo` requires a list of radii to be included in the input.

Listing 9.5: *K*-function (kfunction.py).

```
1  import sys
2  sys.path.append('../geom')
3  from point import *
4  sys.path.append('../indexing')
5  from extent import *
6  from kdtree1 import *
7  from kdtree2b import *
8  import random
9  from math import sqrt, pi
10 import numpy as np
11
12 def kfunc(tree, p, d, density):
13     """
14     Input
15        tree:    a k-D tree
16        p:       a point where the K-function is computed
17        d:       radius of the circle around p
18        density: density of points in the area
19
```

```
    Return
      n: count of points in the circle
      ld: L(d) value
    """
    neighbors = []
    range_query_circular(tree, p, d, neighbors)
    n = len(neighbors)
    kd = n/density
    ld = sqrt(kd/pi)
    return n, ld

def kfunc_monte_carlo(n, area, radii, density, rounds=100):
    """
    Input
      n:      number of points
      area:   Extent object defining the area
      radii:  list containing a set of radii of circles
      rounds: number of simulations
    Return
      percentiles: a list of 2.5th and 97.5th percentiles
                   for each d in radii
    """
    alllds = []
    for test in range(rounds):
      points = [ Point(random.uniform(area.xmin, area.xmax),
                       random.uniform(area.ymin, area.ymax))
                 for i in range(n) ]
      t = kdtree2(points)
      lds = [0 for d in radii]
      for i, d in enumerate(radii):
          for p in points:
              ld = kfunc(t, p, d, density)[1]
              lds[i] += ld
          lds = [ld/n for ld in lds]
          alllds.append(lds)
      alllds = np.array(alllds)
      percentiles = []
      for i in range(len(radii)):
              percentiles.append([np.percentile(alllds[:,i], 2.5),
                                  np.percentile(alllds[:,i], 97.5)])
      return percentiles

def test(points, area):
    """
    Input
      points: a list of Point objects
      area: an Extent object

    Return
      ds: list of radii
```

```
            percentiles: Monte Carlo result of using radii in ds
            lds: L(d) values for each radius in ds
        """
        n = len(points)
        density = n/area.area()
        t = kdtree2(points)
        d = min([area.xmax-area.xmin,area.ymax-area.ymin])*2/3/10
        ds = [ d*(i+1) for i in range(10)]
        lds = [0 for d in ds]
        for i, d in enumerate(ds):
            for p in points:
                ld = kfunc(t, p, d, density)[1]
                lds[i] += ld
        lds = [ld/n for ld in lds]
        percentiles = kfunc_monte_carlo(n, area, ds, density)
        return ds, percentiles, lds

if __name__ == "__main__":
        n = 100
        # random patter
        points = [ Point(random.uniform(20, 30),
                        random.uniform(30, 40))
                   for i in range(n) ]
        xmin = min([p.x for p in points])
        ymin = min([p.y for p in points])
        xmax = max([p.x for p in points])
        ymax = max([p.y for p in points])
        area = Extent(xmin=xmin, xmax=xmax,
                      ymin=ymin, ymax=ymax)
        ds, percentiles, lds = test(points, area)
        print "Random"
        for i, v in enumerate(percentiles):
            print ds[i], v[0], lds[i], v[1]


        # three blocks of points
        points1 = [ Point(random.uniform(10, 20),
                        random.uniform(10, 20))
                   for i in range(n/3) ]
        points2 = [ Point(random.uniform(30, 40),
                        random.uniform(10, 20))
                   for i in range(n/3) ]
        points3 = [ Point(random.uniform(20, 30),
                        random.uniform(30, 40))
                   for i in range(n/3) ]
        points = points1 + points2 + points3
        xmin = min([p.x for p in points])
        ymin = min([p.y for p in points])
        xmax = max([p.x for p in points])
        ymax = max([p.y for p in points])
```

Spatial Pattern and Analysis

```
120         area = Extent(xmin=xmin,xmax=xmax,ymin=ymin,ymax=ymax)
121         ds, percentiles, lds = test(points, area)
122         print "Three blocks"
123         for i, v in enumerate(percentiles):
124             print ds[i], v[0], lds[i], v[1]
```

The test function in Listing 9.5 computes the $L(d)$ values using a series of 10 radii with the maximum radius two thirds of the length of the smaller side of the area (lines 76 and 77). The average value of $L(d)$ is used for each d (line 83). We use two data sets to run the test (Figure 9.4), where the first set of points is randomly distributed and the second has three clusters (but points in each cluster are randomly distributed).

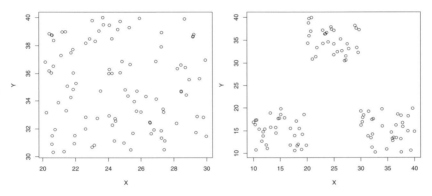

Figure 9.4 Random (left) and clustered (right) points for testing the *K*-function

The outcome is shown below, and we plot a diagonal line, the confidence interval, and the *K*-function values in Figure 9.5. For the random case, it is obvious that the *K*-function values are well within the two lines of the confidence interval, indicating that the random pattern is identified at all scales. The data set with three blocks, however, reveals an interesting dynamic of the spatial pattern. At a small or local scale, meaning when we examine the points near each location, we see a clustered pattern because there appear to be more points than we would expect (considering the size of the area too). However, when the circle radius increases to a certain level (around 10), the pattern starts to appear more and more dispersed as the radius increases. We note that the overall size of each block of points in this case is 10 units, which explains why beyond that scale the overall pattern become dispersed. The example of this case should help to understand the advantages of using the *K*-function: it reveals how spatial pattern is dependent on scale. The nearest neighbor distance statistic does not have this property.

```
Random
0.658746284775  0.782447251926  0.823192147599  0.878923490623
1.31749256955   1.25294384468   1.29708270824   1.41427475067
1.97623885433   1.76164558885   1.80567490505   1.93508213498
2.6349851391    2.25189975889   2.30911526409   2.45740489315
```

```
3.29373142388  2.68641955296  2.81525363907  2.93901601286
3.95247770865  3.07522982843  3.24727346436  3.38607862897
4.61122399343  3.44377836531  3.64967726331  3.82597112471
5.2699702782   3.7986238341   4.00592538201  4.20939096454
5.92871656298  4.1270080257   4.33537220243  4.54396121198
6.58746284775  4.39711802739  4.64973313045  4.82682955853
Three blocks
1.97215401624  2.30732019217  3.35751451614  2.61282902369
3.94430803248  3.80730296142  5.50894499646  4.198217178
5.91646204871  5.32284684915  7.31791426457  5.84865572826
7.88861606495  6.6964392654   8.68481246242  7.37419127158
9.86077008119  8.0705796119   9.56165773421  8.85356120542
11.8329240974  9.31989267977  9.71540908647  10.2497400438
13.8050781137  10.477205022   10.0085157011  11.4893468682
15.7772321299  11.5205533322  10.5218604296  12.6373133211
17.7493861461  12.5309706034  11.2422919216  13.6603894493
19.7215401624  13.4235638982  12.319968568   14.5096243377
```

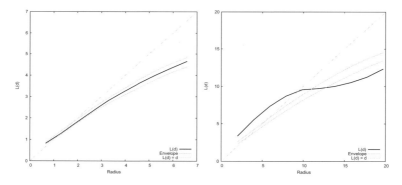

Figure 9.5 The *K*-function values for data sets with random points (left) and with three blocks (right)

Finally, we calculate the *K*-function values for the trees in the Oval using the code in Listing 9.6 and plot the result in Figure 9.6. The result suggests that at a very small scale the *K*-function curve is sandwiched inside the confidence interval. We can compare the *K*-function result with the nearest neighbor distance which indicates a random pattern. Recall that the nearest neighbor distance statistic is calculated using the distance between nearest neighbors, which should fall within the range of small circle radii. It is clear that the *K*-function result tells us more about how the points are organized in space. We should also note that there is a limit to the circle radius. We do not want to endlessly increase the circles. In the code here we impose a two-thirds limit on the length of the smaller side of the area.

Listing 9.6: *K*-function for trees in the Oval (test_oval_trees.py).

```
import sys
sys.path.append('../indexing')
from extent import *
from kfunction import test
from oval_trees import *

area = Extent(xmin=0.0, xmax=6.0, ymin=0.0, ymax=6.0)
ds, percentiles, lds = test(points, area)
for i, v in enumerate(percentiles):
    print ds[i], v[0], lds[i], v[1]
```

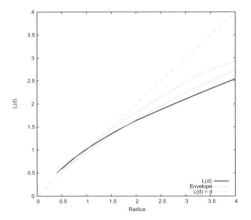

Figure 9.6 The *K*-function values for the trees in the Oval

9.2 Spatial autocorrelation

In the previous section we explored how to analyze patterns of points where we are mainly interested in the location of each point, but not its value. There will be cases when we want to know how the values are spatially related to each other. For example, will we expect an area with high median household income to be spatially close to areas with similar household income values? Or will we see the opposite relationship where high-income areas tend to be more associated with low-income areas? Or will we experience no particular trend, with the high-value areas randomly next to areas with either high or low values? When we must consider continuous values, another set of statistics should be used and Moran's *I* is arguably the most common of these. Here we consider the cross-product of the deviation of the values of a variable at two locations i and j from the mean, and weight it by the proximity between i and j. An average value of all pairs of i and j is then normalized by the variance of the variable for all locations. Formally, we have the following formula for Moran's *I*:

$$I = \frac{N \sum_i \sum_j w_{ij}(z_i - \overline{z})(z_j - \overline{z})}{\sum_i \sum_j w_{ij} \sum_i (z_i - \overline{z})^2},$$

where z_i is the value of the variable at location i, $\overline{z}$ is the mean of the variable among all locations, N is the number of locations, and w_{ij} is an element of a weight matrix assigned to the pair of locations i and j.

The program in Listing 9.7 computes Moran's I when a weight matrix w is given along with the values of each location z. In this program, we basically go through each pair of locations and test if the weight for that pair is non-zero before we calculate the product of the differences. It is straightforward but it seems to focus excessively on pairs that are not adjacent. This way of computing Moran's I will likely be time-consuming.

Listing 9.7: Calculating Moran's I with a weight matrix (moransi.py).

```
1   def moransi(z, w):
2       """
3       Input
4           z: list of values
5           w: weight matrix
6       Output
7           I: Moran's I measure
8       """
9       n = len(z)
10      mean = float(sum(z))/n
11      S0 = 0
12      d = 0
13      var = 0
14      for i in range(n):
15          var += (z[i]-mean)**2
16          for j in range(i):
17              if w[i][j]:
18                  S0 += w[i][j]
19                  d += (z[i]-mean)*(z[j]-mean)
20      I = n*d/S0/var
21      return I
```

When the variable is represented as polygons where each polygon is assigned a set of attribute values, the weight matrix $W = \{w_{ij} \mid 1 \leq i, j \leq N\}$ is often determined by whether polygons are adjacent. In other words, we set $w_{ij} = 1$ if polygons i and j are adjacent, and $w_{ij} = 0$ otherwise. In this case we simplify the weight matrix W into a set that contains only the adjacent pairs, and we can rewrite the Moran's I formula as

$$I = \frac{N \sum_{(i,j) \in W} (z_i - \bar{z})(z_j - \bar{z})}{|W| \sum_i (z_i - \bar{z})^2}.$$

This is how we implement the process of calculating Moran's *I*. A more comprehensive discussion of how to assign weights is given in the exercises.

The program in Listing 9.8 uses the list of adjacent pairs (wlist) to compute Moran's *I*. This should give us a faster method, though we should acknowledge that it takes some time to obtain the list of adjacent pairs.

Listing 9.8: Calculating Moran's *I* with an adjacent list (moransi2.py).

```
def moransi2(z, wlist):
    """
    Input
       z: list of values
       w: weight list
    Output
       I: Moran's I measure
    """
    n = len(z)
    d = 0.0
    var = 0.0
    mean = float(sum(z))/n
    for i in range(n):
        var += (z[i]-mean)**2
    S0 = len(wlist)
    for e in wlist:
        d += (z[e[0]]-mean)*(z[e[1]]-mean)
    I = n*d/S0/var
    return I
```

We will test the Moran's *I* code in Listing 9.8 using the population density in the 3,109 conterminous counties of the United States in the 2000 census data. In the code in Listing 9.9 we use the adjacency matrix created in Appendix B.1.5, where the matrix is saved in the Python pickle format. We load the pickle file directly back into the program. We also convert the adjacent matrix into a list (line 27) and use that to calculate Moran's *I*.

Listing 9.9: Testing Moran's *I* using a weight matrix (test_moransi.py).

```
from osgeo import ogr
from numpy import *
import pickle
import time
```

```
5   from moransi import *
6   from moransi2 import *
7
8   driver = ogr.GetDriverByName("ESRI Shapefile")
9   fname = '../data/uscnty48area.shp'
10  vector = driver.Open(fname, 0)
11  layer = vector.GetLayer(0)
12  n = layer.GetFeatureCount()
13
14  adj = pickle.load(open('../gdal/uscnty48area.adj.pickle'))
15
16  z = []
17  for i in range(n):
18      feature1 = layer.GetFeature(i)
19      z.append(feature1.GetField("PopDensity"))
20
21  t1 = time.time()
22  I = moransi(z, adj)
23  t2 = time.time()
24  print "I =", I
25  print "Time =", t2-t1
26
27  wlist = []
28  for i in range(n):
29      for j in range(i):
30          if adj[i][j]:
31              wlist.append([i, j])
32
33  t3 = time.time()
34  I = moransi2(z, wlist)
35  t4 = time.time()
36  print "I =", I
37  print "Time =", t4-t3
```

The result of running the above code is given below. We observe a stunning difference in computing time between the two methods. We will see later how this will make a huge difference in evaluating spatial autocorrelation.

```
I = 0.239939848606
Time = 2.76858615875
I = 0.239939848606
Time = 0.00218296051025
```

When the spatial pattern is random, the expected value of its Moran's I is

$$E[I] = -\frac{1}{N-1}$$

and the variance is

$$\text{var}(I) = \frac{N^2 S_1 - N S_2 + 3 S_0^2}{(N^2-1) S_0^2} - E[I]^2$$

$$= \frac{N^2(N-1)S_1 - N(N-1)S_2 + 2(N-2)S_0^2}{(N+1)(N-1)^2 S_0^2},$$

where

$$S_0 = \sum_i \sum_j w_{ij},$$

$$S_1 = \frac{1}{2} \sum_i \sum_j (w_{ij} + w_{ji})^2,$$

$$S_2 = \sum_k \left(\sum_j w_{kj} + \sum_i w_{ik} \right)^2.$$

The above formula for the variance of Moran's I, however, is derived only when the values are independently drawn from a normal distribution. Another kind of random pattern can be achieved when the values are shuffled among the locations. In this case, we have a different formula for the variance:

$$\text{var}(I) = \frac{N S_3 - S_4 S_5}{(N-1)(N-2)(N-3) S_0^2} - E[I]^2,$$

where

$$S_3 = (N^2 - 3N + 3) S_1 - N S_2 + 3 S_0^2,$$

$$S_4 = \frac{\frac{1}{N} \sum_i (z_i - \bar{z})^4}{[\frac{1}{N} \sum_i (z_i - \bar{z})^2]^2},$$

$$S_5 = N(N-1) S_1 - 2 N S_2 + 6 S_0^2.$$

The variances are certainly computable, though tedious. We do not directly implement them here, but leave that to the exercises. Instead, we will use the same approach as we did with the K-function: Monte Carlo simulation. The idea here is to generate a large number of random patterns and compute Moran's I for each of them. Then we compare the index we obtain from the original data against the simulations and determine whether the index we have is significantly different from the simulation samples. The entire code should be straightforward (Listing 9.10). We compute 10,000 simulations here, which takes about half a minute. It would be impractical to use the matrix version of Moran's *I* unless some computing tricks can dramatically increase the efficiency. The last part of the code uses the plotting tools in Python to create a graph.

Listing 9.10: Monte Carlo simulation on Moran's *I* (test_moransi_mc.py).

```
1   from osgeo import ogr
2   from numpy import *
3   import pickle
4   import time
5   from random import shuffle
6   from moransi2 import *
7   import matplotlib.pyplot as plt
8
9   driver = ogr.GetDriverByName("ESRI Shapefile")
10  fname = '../data/uscnty48area.shp'
11  vector = driver.Open(fname, 0)
12  layer = vector.GetLayer(0)
13  n = layer.GetFeatureCount()
14
15  adj = pickle.load(open('../gdal/uscnty48area.adj.pickle'))
16  z = []
17  for i in range(n):
18      feature1 = layer.GetFeature(i)
19      z.append(feature1.GetField("PopDensity"))
20
21  I = moransi2(z, wlist)
22  print I
23  wlist = []
24  for i in range(n):
25      for j in range(i):
26          if adj[i][j]:
27              wlist.append([i, j])
28
29  ms = []
30  for i in range(10000):
31      shuffle(z)
32      ms.append(moransi2(z, wlist))
33
34  print percentile(ms, 97.5)
35  fig = plt.figure()
36  ax = fig.add_subplot(1, 1, 1)
37  ax.hist(ms, 20, color="grey", alpha=0.8)
38  plt.show()
39  fig.savefig('moransi-mc.eps')
```

The histogram of the Monte Carlo simulation is shown in Figure 9.7. The 97.5th percentile is 0.0223. It is clear that the value of Moran's I for our data (0.2399) is well beyond that range and we can therefore conclude that there is a significant positive autocorrelation in the population density data of our map.

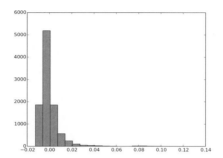

Figure 9.7 Histogram of Monte Carlo simulations of Moran's I values

Finally, we test a different spatial data set on a controlled grid (Listing 9.11). We use the list of adjacent pairs to compute the Moran's I values. But the actual code for creating this list is not directly given in this book. Instead, it is one of the exercises set at the end of this chapter (Exercise 4). We now assume that the code exists and returns a list of adjacent pairs as expected (lines 3 and 11). The grid has 15 rows and 15 columns as specified in line 9. Line 15 creates a random set of independent values in the grid and the next line computes the value of Moran's I for the data. We repeat this 10,000 times. In a second scenario, we stick with one particular set of random data and shuffle it 10,000 times (line 22). In this sense, we are testing cases where the values are permuted. In a third scenario, we create a special pattern of values where the highest value of 1 is at the center of the grid and from there the values decrease in accordance with the formula $\frac{15}{16}[1-(d/r)^2]^2$, where d is the distance from a location in the grid to the center and r is called the bandwidth. Here we set the bandwidth to be the size of the grid ($r = 15$). The formula is a commonly used distance decay kernel function, also known as the biweight kernel.

Listing 9.11: Testing Moran's I using data on a regular grid (test_moransi_grid.py).

```
1  from numpy import percentile
2  from moransi2 import *
3  from create_wlist import create_wlist  # TO DO
4  from random import shuffle
5  from random import random
6  from math import sqrt
7  import matplotlib.pyplot as plt
8
9  n1 = 15
10 n = n1*n1
11 wlist = create_wlist(n1, n1)  # TO DO
```

```
12
13   ms1 = []
14   for i in range(10000):
15       data = [random() for i in range(n)]
16       I = moransi2(data, wlist)
17       ms1.append(I)
18
19   ms2 = []
20   data = [random() for i in range(n)]
21   for i in range(10000):
22       shuffle(data)
23       I = moransi2(data, wlist)
24       ms2.append(I)
25
26   i0 = n1/2
27   j0 = n1/2
28       for i in range(n1):
29           for j in range(n1):
30               d = sqrt((i-i0)*(i-i0) + (j-j0)*(j-j0))
31               val = 15.0/16.0 * (1 - (d/n1)**2)**2
32               pos = i*n1+j
33               data[pos] = val
34
35   I = moransi2(data, wlist)
36   print I
37
38   print percentile(ms1, 2.5), percentile(ms1, 50),\
39       percentile(ms1, 97.5)
40   print percentile(ms2, 2.5), percentile(ms2, 50),\
41       percentile(ms2, 97.5)
42
43   fig = plt.figure()
44   ax = fig.add_subplot(1, 1, 1)
45   ax.hist(ms1, 20, color="white",alpha=0.8,label="Independent")
46   ax.hist(ms2, 20, color="grey",alpha=0.3,label="Permutation")
47   plt.legend(loc='upper right')
48   plt.show()
```

The output of the code is as follows:

```
0.879154984566
-0.0983709874209 -0.00442892038325 0.0918736611396
-0.0998426464734 -0.00425357474198 0.0909170028323
```

It is obvious that the distance decay kernel has a high positive autocorrelation value and the random patterns all have Moran's I value close to -0.004 (this is very close to the theoretical mean of $-1/(N-1) = -0.00446$). The shapes of the two histograms are close as well, though one may argue that the independent case has a more concentrated distribution (Figure 9.8).

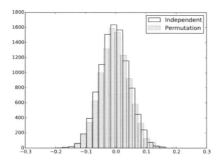

Figure 9.8   Histogram of Monte Carlo simulations of Moran's *I* values on regular grids

9.3 Clustering

A consequence of positive spatial autocorrelation, as it appears in spatial patterns, is the emergence of clusters, meaning that related things are close to each other and form clusters. There are many ways to identify where those clusters are. Here we introduce a commonly used algorithm called the *k*-means algorithm. To use this algorithm, we first need to tell it how many clusters there are in the data, which is indicated by an integer *k*. The algorithm first places *k* points randomly in the area, and these points are used as the initial center points of the clusters; we will call each of them a center. Then we assign each point in the data to its nearest center. The points assigned to each center are used to compute their own center, which can be different from the center point to which they have been assigned. If there is a significant difference between the new and old centers, we will use the new centers to replace the old ones and redo the assignment. This will yield a gain in total distance. We repeat the process until there is no significant gain in total distance. The *k*-means algorithm is surprisingly simple and flexible in the sense that the original data set does not have to be spatial at all. It can be any data in any dimension. It is one of the most commonly used algorithms in the fields of machine learning and data mining.

We implement the *k*-means algorithm in Listing 9.12. Specifically, we use the function `clustering_dist` to assign each point to its nearest center and to return the total distance after the assignment. The function `initk` returns an initial set of centers. There are different initialization methods, and here we implement two of them. The first one is called the Forgy method and randomly chooses *k* points from the input points as the seed centers, while the random method uses *k* random locations generated using a random number generator.

Listing 9.12: Clustering using *k*-means (kmeans.py).

```
import sys
sys.path.append('../geom')
from point import *
```

```
 4  import random
 5  from math import fabs
 6  from math import sqrt
 7
 8  INF = float('inf')
 9
10  def clustering_dist(points, means):
11      n = len(points)
12      k = len(means)
13      nearests = [[] for i in range(k)]
14      totaldist = 0
15      for i in range(n):
16          dmin = INF
17          near = []
18          for j in range(k):
19              d = points[i].distance(means[j])
20              if d < dmin:
21                  dmin = d
22                  jmin = j
23          totaldist += dmin
24          nearests[jmin].append(i)
25      totaldist = totaldist/n
26      return nearests, totaldist
27
28  def initk(points, k, init):
29      n = len(points)
30      xmin = INF
31      ymin = INF
32      xmax = -INF
33      ymax = -INF
34      for p in points:
35          xmin = min([xmin, p.x])
36          ymin = min([ymin, p.y])
37          xmax = max([xmax, p.x])
38          ymax = max([ymax, p.y])
39      nearests = [[] for i in range(k)]
40      while [] in nearests:       # until no empty set in nearests
41          if init=="forgy":       # Forgy initialization
42              means = [points[i]
43                       for i in random.sample(range(n), k)]
44          elif init=="random":
45              means = [ Point(random.uniform(xmin, xmax),
46                              random.uniform(ymin, ymax))
47                       for i in range(k) ]
48          else:
49              print "Error: unknown initialization method"
50              sys.exit(1)
51          nearests, totaldist = clustering_dist(points, means)
52      return means, nearests, totaldist
53
54  def kmeans(points, k, threshold=1e-5, init="forgy"):
```

```python
        bigdiff = True
        means, nearests, totaldist = initk(points, k, init)
        while bigdiff:
            means2 = []
            for j in range(k):
                cluster = [xx for xx in nearests[j]]
                sumx = sum([points[ii].x for ii in cluster])
                sumy = sum([points[ii].y for ii in cluster])
                numpts = len(nearests[j])
                if numpts>0:
                    sumx = sumx/numpts
                    sumy = sumy/numpts
                means2.append(Point(sumx, sumy))
            nearests, newtotal = clustering_dist(points, means2)
            offset = totaldist - newtotal
            if offset > threshold:
                means = means2
                totaldist = newtotal
            else:
                bigdiff = False
        return totaldist, means

def test():
    n = 500
    points = [ Point(random.random(), random.random())
               for i in range(n) ]
    print kmeans(points, 10, init="forgy")[0]

    points1 = [ Point(random.uniform(10, 20),
                      random.uniform(10, 20))
                for i in range(n/2) ]
    points2 = [ Point(random.uniform(30, 40),
                      random.uniform(30, 40))
                for i in range(n/2) ]
    print kmeans(points1+points2, 2)[0]

    points1 = [ Point(random.uniform(10, 20),
                      random.uniform(10, 20))
                for i in range(n/3) ]
    points2 = [ Point(random.uniform(30, 40),
                      random.uniform(10, 20))
                for i in range(n/3) ]
    points3 = [ Point(random.uniform(20, 30),
                      random.uniform(30, 40))
                for i in range(n/3) ]
    print kmeans(points1+points2+points3, 3)[0]
    print kmeans(points1+points2+points3, 3,
                 init="random")[0]

if __name__ == "__main__":
    test()
```

In the *k*-means code, we also test three scenarios with 500 points. The first case has the points randomly distributed. The second and third cases have two and three blocks, respectively, where points in each block are randomly distributed. Below is a run of these cases, where the last two lines are identical, resulting from different initialization methods for the third case. Figure 9.9 shows how the center points move toward the final result using numbers to indicate the iterations.

```
0.118536697855
3.77665952559
3.86220805887
3.86220805887
```

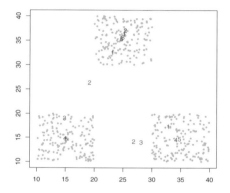

Figure 9.9 Example run of the *k*-means algorithm. The numbers indicate the iterations of the program run where 1s indicate the initial centers

9.4 Landscape ecology metrics

The main purpose of studying any spatial pattern is that we hope the pattern or the appearance of a spatial phenomenon can lead us to an understanding of the process that causes the pattern. For example, having concluded that certain locations are clustered, we may well ask why this is so. This is indeed one of the most important aspects of landscape ecology, which studies patterns and more importantly how patterns are measured. There are many different ways to assess the patterns of landscape and we will of course not be able to cover all of them. Instead, we will focus on one that can serve as a good start point. Before we begin, however, it is necessary to mention that most landscape ecology metrics are applied to categorical data. Raster data are commonly used in this context, though some landscape ecology metrics are measured on vector data sets.

Contagion is a landscape ecology metric that measures the probability of having a different type at each location. A high contagion value indicates a generally clumped pattern of types where one class of land takes up a big portion of the entire area, while a low contagion value indicates a dispersed and fragmented pattern.

Let p_{ij} be the probability of two randomly chosen adjacent pixels belonging to types i and j, respectively. We can estimate p_{ij} as

$$p_{ij} = p_i N_{ij} / N_i,$$

where p_{ij} is the probability of a randomly chosen pixel being of type i, N_i is the total number of occurrences of type i being on one side of an edge, and N_{ij} is the number of adjacencies where types i and j are on both sides. The contagion measure of the landscape can be calculated using an entropy-like formula,

$$C = 1 + \frac{\sum_i^n \sum_j^n p_{ij} \ln p_{ij}}{2 \ln n},$$

where n is the number of types in the data.

We implement the calculation of the contagion metric in Python (Listing 9.13). The input must be a two-dimensional array of integers because the metric only applies to categorical data. At the beginning of the program we need to identify the unique values in the data and store those values in a list (line 16). The proportion of each type (p_i) is calculated in line 27. We then use a Python dictionary data structure (types) to sequentially index each of the unique values, as the integers in the data may not be continuous (line 29). The rest of the code is straightforward: we check each edge between cells in the data, in columns and rows, and count the number of edges between different pairs of types. We do not count the edges with the same type on both sides because we focus on patches, not pixels, in this metric. With the counts, we can calculate the value of p_{ij} (line 61), which can then be used to compute the contagion measure.

Listing 9.13: Calculation of contagion (contagion.py).

```
1  import numpy as np
2  from sys import exit
3  from math import log
4
5  def contagion(data):
6      """
7      Input
8          data: a NumPy 2-dimensional array of integers
9
10     Output
11         Contagion metric
12     """
13     nrows = len(data)
14     ncols = len(data[0])
15     # get the list of unique values
16     UList = []
```

```
        Pi = []
        for i in range(nrows):
            for j in range(ncols):
                if data[i, j] not in UList:
                    UList.append(data[i, j])
                    Pi.append(1)
                else:
                    itype = UList.index(data[i,j])
                    Pi[itype] += 1

        Pi = [float(i)/ncols/nrows for i in Pi]

        types = dict()
        for i in range(len(UList)):
            types[UList[i]] = i

        n = len(UList)
        Nij=[[0]*n for x in range(n)]
        Ni = [0 for x in range(n)]
        for i in range(nrows):              # edges in rows
            for j in range(ncols-1):
                i1 = types[data[i, j]]
                j1 = types[data[i, j+1]]
                Ni[i1] += 1                 # type i1 on one side
                Ni[j1] += 1                 # j1 on the other side
                if i1==j1:
                    continue
                Nij[i1][j1] += 1
                Nij[j1][i1] += 1            # symmetrical
        for j in range(ncols):              # edges in columns
            for i in range(nrows-1):
                i1 = types[data[i,j]]
                j1 = types[data[i+1, j]]
                Ni[i1] += 1
                Ni[j1] += 1
                if i1==j1:
                    continue
                Nij[i1][j1] += 1
                Nij[j1][i1] += 1            # symmetrical

        sum = 0.0
        for i in range(n):
            for j in range(n):
                if Nij[i][j]:
                    pij = float(Nij[i][j]) * Pi[i] / Ni[i]
                    sum += pij * log(pij)

        sum /= 2.0*log(n)
        sum += 1
        return sum
```

Spatial Pattern and Analysis

We first test the contagion metric using a few simple spatial configurations on a 4 x 4 grid (Listing 9.14). These data include a pattern with four types where each type is only next to a different type (data1), a similar pattern but with only two types in a checkerboard format (data2), a pattern with one relatively dominant type but otherwise fragmented (data3), a four-type pattern with four equal blocks (data4), and a pattern that is dominated by a big patch (data5).

Listing 9.14: Testing of contagion, part 1 (test_contagion.py).

```
import random
from contagion import *
import numpy as np

data1 = np.array([[ 1, 2, 3, 4],
                  [ 4, 3, 2, 1],
                  [ 1, 2, 3, 4],
                  [ 4, 3, 2, 1]])

data2 = np.array([[ 1, 0, 1, 0],
                  [ 0, 1, 0, 1],
                  [ 1, 0, 1, 0],
                  [ 0, 1, 0, 1]])

data3 = np.array([[ 0, 1, 1, 1],
                  [ 1, 1, 1, 1],
                  [ 0, 1, 1, 1],
                  [ 1, 1, 0, 0]])

data4 = np.array([[1, 1, 2, 2],
                  [1, 1, 2, 2],
                  [3, 3, 4, 4],
                  [3, 3, 4, 4]])

data5 = np.array([[ 1, 1, 1, 1],
                  [ 1, 1, 1, 1],
                  [ 1, 1, 1, 1],
                  [ 1, 1, 0, 0]])
print "data1:"
print data1, contagion(data1)
print "data2:"
print data2, contagion(data2)
print "data3:"
print data3, contagion(data3)
print "data4:"
print data4, contagion(data4)
print "data5:"
print data5, contagion(data5)
```

The results of these five cases are given below. The results are consistent with the design principle of contagion as we can examine the tradeoff effect between fragmentation (`data1`) that tends to decrease the contagion measure and patch size (`data5`) that increases it.

```
data1:
[[1 2 3 4]
 [4 3 2 1]
 [1 2 3 4]
 [4 3 2 1]]  0.270741104622
data2:
[[1 0 1 0]
 [0 1 0 1]
 [1 0 1 0]
 [0 1 0 1]]  0.5
data3:
[[0 1 1 1]
 [1 1 1 1]
 [0 1 1 1]
 [1 1 0 0]]  0.557573110559
data4:
[[1 1 2 2]
 [1 1 2 2]
 [3 3 4 4]
 [3 3 4 4]]  0.617919791607
data5:
[[1 1 1 1]
 [1 1 1 1]
 [1 1 1 1]
 [1 1 0 0]]  0.736734583866
```

While the above experiments give us a casual inspection of factors that affect contagion, here we carry out a more extensive examination. In Listing 9.15 we utilize the idea of a kernel function to create a patch at the center of the grid. Four kernel functions are implemented here: biweight, triweight, Gaussian, and uniform (lines 29, 31, 33, and 35, respectively). We test the default biweight kernel. All cells with a value greater than 0.3 will be assigned a type of 3 (line 37) while all other pixels will be randomly assigned a value between 0 and 2 (line 39). The size of the center patch is controlled by a bandwidth, and we test the impact of a range of bandwidths from 1.1 to 14.5 in steps of 0.5 (line 45). We plot the relationship between contagion and bandwidth in Figure 9.10, where it is obvious that contagion increases with the bandwidth value that controls the size of the center patch.

Listing 9.15: Testing of contagion, part 2 (test_contagion_kernel.py).

```python
import random
from contagion import *
from math import sqrt, pi, exp
import numpy as np

def kernel(nrows, ncols, r, func="biweight"):
    """
    Returns a grid of cells with a core at the center.
    Input
      nrows: number of rows
      ncols: number of columns
      r: the bandwidth of the kernel
      func: kernel function (biweight, triweight,
            gaussian, or uniform)
    Output
      data: a two-dimensional list (of lists)
    """
    data = np.array([ [0]*ncols for i in range(nrows)])
    i0 = nrows/2
    j0 = ncols/2
    for i in range(nrows):
        for j in range(ncols):
            d = sqrt((i-i0)*(i-i0) + (j-j0)*(j-j0))
            if d>r:
                val = 0
            else:
                u = float(d)/r
                if func=="biweight":
                    val = 15.0/16.0 * (1 - u**2)**2
                elif func=="triweight":
                    val = 35.0/32.0 * (1 - u**2)**3
                elif func=="gaussian":
                    val = 1.0/sqrt(2*pi) * exp(-0.5*u*u)
                else:
                    val = 0.5
            if val > 0.3:
                val = 3
            else:
                val = random.randint(0, 2)
            data[i][j] = val
    return data

nrows = 15
ncols = 15
rs = np.arange(1.1, ncols, 0.5)
for r in rs:
    data = kernel(nrows, ncols, r, func="biweight")
    print r, contagion(data)
```

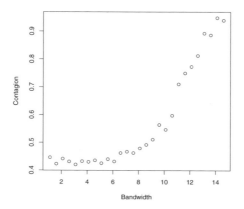

Figure 9.10 Contagion

9.5 Notes

The quantitative geography literature is full of discussions on spatial pattern analysis (Cliff and Ord, 1981; Fotheringham et al., 2000). The book by Boots and Getis (1988) provides a good summary of many point pattern analysis techniques. The book by Bailey and Gatrell (1995) gives an in-depth exploration of more methods. The nearest neighbor statistic was developed by Clark and Evans (1954), and the K-function was developed by Ripley (1981).

Although Moran's index (Moran, 1950) has been in widespread use, there are other types of statistics such as Geary's C (Geary, 1954) that can provide indications about spatial autocorrelation. Here we mention local statistics, a notable new development in the measurement of spatial autocorrelation. For example, the local Moran's I statistic (Anselin, 1995) and the G and G^* statistics (Getis and Ord, 1992) extend the idea to each location so that it is possible to examine the spatial variation of spatial autocorrelation. We do not directly cover these new developments in this chapter. However, studying topics such as spatial adjacency and Moran's I should equip the reader to explore new directions. The contagion metric was first developed by O'Neill et al. (1988). The initial formula, however, does not normalize the final result, and a corrected version provides more reliable results for interpretation (Riitters et al., 1996).

Although we use Python to show the computation of spatial pattern analysis in this chapter (and this book in general), there is a rich collection of software packages designed for spatial analysis purposes. The software package called FRAGSTATS[2] (McGarigal and Marks, 1995) provides a comprehensive set of tools to measure landscape patterns using a large number of metrics. A pure Python library called PySAL[3] has tools for advanced (and fundamental) spatial analysis needs. There are many libraries

---

[2] http://www.umass.edu/landeco/research/fragstats/fragstats.html

[3] http://pysal.org

in R, among them sdep.[4] In Appendix B.3 we give a very short tutorial on PySAL for the calculation of Moran's *I*.

9.6 Exercises

1. An advantage of the nearest neighbor distance statistic is that it is relatively easy to explain because we only need to deal with one value. In contrast, the *K*-function varies a lot when we compute it from different points in the study area. But a nice feature of the *K*-function is that it gives us a sense of scale: we can tell how the clustering or dispersion pattern varies when we examine it from different scales (radii). For the nearest neighbor distance measure, one must wonder why it has to be the *nearest* neighbor distance. Why not the second nearest neighbor distance? With our powerful indexing tool, we can easily explore further in this direction. Design a program that returns a distance measure as a function of the radius and explain how this can be used to obtain more insight into the pattern of a spatial phenomenon. We can try, for example, the average distance from a point to other points in a given radius.

2. We used the *k*-D tree to speed up the calculation of the *K*-function. Examine what happens if we do not use any indexing method.

3. Compare different indexing methods and see how they perform in the specific task of computing the *K*-function for different data sets.

4. Computing an adjacency matrix for a vector data set is often tedious. For a raster data set where the spatial units are regular squares, however, it seems there is no such need because the units are regularly organized and the adjacency between two units (cells) can be easily determined by knowing where the two units are. Write a Python function called `create_wlist` to do just that and then plug it into the Moran's *I* code to test how this works. The function takes two input arguments: the number of rows and columns, in that order. It returns a list where each element is another list of two integers indicating the indices of two adjacent units.

5. Moran's *I* is not the only statistic that can be used to test spatial autocorrelation. Another index is called Geary's *C*, given by

$$C = \frac{(n-1)\sum_i \sum_j w_{ij}(z_i - z_j)^2}{2\sum_i \sum_j w_{ij}(z_i - \bar{z})^2}.$$

Write a Python program to compute Geary's *C* for any spatial data. Compare Geary's *C* and Moran's *I* for the same data set.

---
[4] http://cran.r-project.org/web/packages/spdep/index.html

6. The weight matrix used in our Moran's I computation is based on polygon adjacency. This may seem arbitrary since we can certainly come up with other ways of determining whether two locations (polygons in our case) are close enough to be considered in the calculation. For example, it is possible to use an internal point of each polygon (e.g., the centroid), and the two polygons have a weight of 1 if their internal points are within a distance threshold. We can also force the program to only consider up to k close locations. We use a binary weight in this book. But the weight can also be continuous, proportional to, say, inverse distance or the square of inverse distance. Implement some of these ideas and compare the new indices with Moran's I.

7. Write a Python program to calculate the two variances of Moran's I and compare the hypothesis tests using these variances with the Monte Carlo simulation results.

10

Network Analysis

> The Road goes ever on and on
> Down from the door where it began.
> Now far ahead the Road has gone,
> And I must follow, if I can,
> Pursuing it with eager feet,
> Until it joins some larger way
> Where many paths and errands meet.
> And whither then? I cannot say.
>
> J. R. R. Tolkien, *The Fellowship of the Ring*

Networks are ubiquitous. Transportation, for example, relies on the existence of road networks to be functional. Computer networks connect between different computing devices. Social media applications require an effective network of people to be useful. While different kinds of networks in reality have different features, we can understand and process them using two fundamental concepts: nodes and edges. Nodes represent functional units on the networks that act as the generating and receiving sides of functions, information, services, and goods. Edges represent the media enabling functions, information, services, and goods to move from one node to another. Both nodes and edges can be weighted to suit different applications. For example, in a transportation network, nodes can be weighted as the total amount of goods to be moved and edges as the capacity and/or length. Social networks may be weighted so that each node is a person (and probably everything that comes with that person in the cyber world) and each edge represents some form of relationship between people (such as different kinds of friendship). Networks can be further conceptualized as graphs where we are not as interested in how nodes and edges are weighted as we are in what can we do with the nodes and edges.

As with other spatial data types, networks need a good representation scheme so that we can effectively process them in the computer. Let us take the simple network illustrated in Figure 10.1 as an example. Here we have eight nodes labeled with italic numbers, and edges between nodes that are directly connected. The number on each edge represents some kind of cost measure (weight) that may be distance or the time required to go from one node to an immediately connected one. There are two general approaches that can be used to encode this network. The first is a list of lists where each element in the list refers to a node and the element contains another list that includes all the nodes adjacent to the element. Additional information such as the weight on each edge can also be added into this data structure. This can be stored in a simple text file and loaded into any program for further processing. This data/file structure is simple but does not give us information about the connections between nodes that are not adjacent. The second data structure has the potential to solve this problem using a matrix where the value at each cell is the total weight for the shortest connection between the two nodes indicated by the row and column of the cell. The matrix representation, ultimately equivalent to the list, requires us to know the shortest connections between all node pairs, which typically do not come immediately when we have the network data.

Typically, we start building network data more simply by listing each pair of nodes that are adjacent (Listing 10.1). This simple node–edge view of the network can be used in a list structure. It can also be used to build a matrix for the network after some computing. This is really the main focus of this chapter: we will introduce the algorithms that can be used to derive the shortest paths between network nodes. While network analysis can mean many things, these algorithms are designed to tackle the most fundamental tasks on networks and should serve as a great start point on which to build. We will use our simple network in Figure 10.1 to test all the algorithms. It should be noted that we only discuss undirected networks in this book.

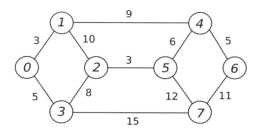

Figure 10.1 An example network with nodes labeled with italic numbers and edges marked with their lengths (not to scale)

Listing 10.1: List storing information about each edge for the simple network in Figure 10.1 (network-links.dat).

```
8
0 1 3
0 3 5
1 2 10
1 4 9
2 3 8
2 5 3
3 7 15
4 5 6
4 6 5
5 7 12
6 7 11
```

The code in Listing 10.2 includes two functions. The `network2list` function reads a network file (see Listing 10.1) into a list of lists to implement the network list view as discussed above. The `network2distancematrix` function converts the file into a matrix form. Here, for the matrix format, we only record the distance or weight measure for the immediate connections (i.e., edges) while using an infinity value (INF) for all the other connections between nodes. Later in this chapter when we introduce the all-pair shortest distance algorithm we will complete the puzzle by replacing the INF values in the matrix. These two functions, especially `network2distancematrix`, will be frequently used in this and the next two chapters where networks are discussed.

Listing 10.2: Reading a network file into a list or matrix (network2listmatrix.py).

```python
INF = float('inf')

def network2list(fname, is_zero_based = True):
    """
    Converts a network file to a list data structure.
    Input
      fname: name of a network file in the following format
             n
             i j distance
             ...
    Output
      network: a list of lists where the i-th element is the
               list of all nodes adjacent to node i
    """
    f = open(fname)
    l = f.readline()
    n = int(l.strip().split()[0])
    network = [[] for i in range(n)]
    for l in f:
        nodesnedge = l.strip().split()
        if len(nodesnedge)==3:
            i=int(nodesnedge[0])
            j=int(nodesnedge[1])
            if not is_zero_based:
                i = i-1
                j = j-1
            network[i].append(j)
            network[j].append(i)
    f.close()
    return network

def network2distancematrix(fname, is_zero_based = True):
    """
    Reads a list from the input graph file and returns
    the adjacent matrix. The input file has n+1 lines and
    must be in this format:

    n
    i j distance
```

```
40            ...
41
42            where the first line has the number of edges,
43            and each of the remaining lines have the indices and
44            distance on each edge. Indices start from 0 (default)
45            or 1.
46            """
47            a = []
48            f = open(fname)
49            l = f.readline()
50            n = int(l.strip().split()[0])      # number of nodes
51            a=[[INF]*n for x in xrange(n)]    # init 2D list of INF
52            for l in f:
53                nodesnedge = l.strip().split()
54                if len(nodesnedge)==3:
55                    i=int(nodesnedge[0])
56                    j=int(nodesnedge[1])
57                    if not is_zero_based:
58                        i = i-1
59                        j = j-1
60                    d=int(nodesnedge[2])
61                    a[i][j] = d
62                    a[j][i] = d
63            for i in range(n):
64                a[i][i] = 0
65            return a
```

10.1 Network traversals

Starting from any node in a network, what is the most effective way to go through all the other nodes based on the connectivity between those nodes? This is a problem of network traversal, which is different from simply enumerating the nodes without any structure. Here we want to make sure the nodes are only visited because of how they connect with each other. There are two widely used methods for this purpose: breadth-first and depth-first traversal.

10.1.1 Breadth-first traversal

Given a start node, we can visit all the other nodes in a network by first visiting the nodes that are adjacent to the given node, and then starting from each adjacent node we do the same by visiting its adjacent nodes. This process repeats until there are no more nodes to visit. Throughout the process, we insert each node visited or to be visited into a data structure and we make sure that the nodes inserted first will be popped out (removed) first. This data structure is a queue that enforces a first-in-first-out (FIFO) rule. Because in this way we always search for the nodes near each node, it is a breadth-first search approach.

We implement the breadth-first search in Listing 10.3. Here we simply use a Python list as a queue by ensuring new items are always added to the end of the list (line 14) and removal is always done at the beginning of the list (line 20). The queue in this algorithm holds the nodes that we will visit, and we dynamically maintain the queue so that in each iteration we pop out the first node in the queue (line 20). Meanwhile, we maintain two additional data structures to keep all the nodes that are visited in list V (line 21) and a list called `labeled` where we mark each node as `True` if it has been added to the queue (line 16). We find all the nodes adjacent to the one just popped out, and if the adjacent node has not been considered before (line 23) we will label the node (line 24) and append it to the queue (line 25).

Listing 10.3: Breadth-first search on a network (bfs.py).

```
1   from network2listmatrix import network2list
2
3   def bfs(network, v):
4       """
5       Breadth-first search
6       Input
7         network: a network represented using a list of lists
8         v: initial node to start search
9       Output
10        V: a list of nodes visited
11      """
12      n = len(network)
13      Q = []
14      Q.append(v)
15      V = []
16      labeled = [ False for i in range(n) ]
17      labeled[v] = True
18      while len(Q) > 0:
19          # print list(Q)
20          t = Q.pop(0)
21          V.append(t)
22          for u in network[t]:
23              if not labeled[u]:
24                  labeled[u] = True
25                  Q.append(u)
26      return V
27
28  if __name__ == "__main__":
29      network = network2list('../data/network-links')
30      V = bfs(network, 3)
31      print "Visited:", V
```

We test the breadth-first search algorithm using our simple network and an initial node of 3. Below we also print out the contents in the queue during each iteration and the final sequence of nodes being visited.

```
[3]
[0, 2, 7]
[2, 7, 1]
[7, 1, 5]
[1, 5, 6]
[5, 6, 4]
[6, 4]
[4]
Visited: [3, 0, 2, 7, 1, 5, 6, 4]
```

Note that the queue constantly changes its content by maintaining the nodes adjacent to the node that is previously the first in the queue and is popped in each iteration. The final sequence of nodes is formed by the first node in the list during each iteration.

10.1.2 Depth-first traversal

The depth-first search approach treats adjacent nodes differently than the breadth-first search because now we go on with one adjacent node and keep searching for the nodes adjacent to that node until we cannot continue. In order to do so, we will use a different data structure called a stack that only allows the most recently added items to pop out (so we call this last-in-first-out, or LIFO). This is implemented in Listing 10.4 where we use the Python list to realize the functions of a stack (line 13). Other than that, the overall structure of the algorithm is almost identical to that in the breadth-first search.

Listing 10.4: Depth-first search on a network (dfs.py).

```
1   from network2listmatrix import network2list
2
3   def dfs(network, v):
4       """
5       Depth-first search
6       Input
7          network: a network represented using a list of lists
8          v: initial node to start search
9       Output
10         V: a list of nodes visited
11      """
12      n = len(network)
13      S = [] # empty stack
14      S.append(v)
15      V = []
16      labelled = [ False for i in range(n)]
17      labelled[v] = True
18      while len(S) > 0:
19          print S
20          t = S.pop()
21          V.append(t)
```

```
22                  for u in network[t]:
23                      if not labelled[u]:
24                          labelled[u] = True
25                          S.append(u)
26              return V
27
28      if __name__ == "__main__":
29          network = network2list('../data/network-links')
30          V = dfs(network, 3)
31          print "Visited:", V
```

We again test the algorithm using the simple network and node 3 as the initial node:

```
[3]
[0, 2, 7]
[0, 2, 5, 6]
[0, 2, 5, 4]
[0, 2, 5, 1]
[0, 2, 5]
[0, 2]
[0]
Visited: [3, 7, 6, 4, 1, 5, 2, 0]
```

The output from testing the depth-first search clearly demonstrates the difference between the two traversal algorithms. Here we can see that, after the initial step where we insert the first node, the first element in the stack (0) is never changed until the last iteration is complete when the stack becomes empty. This is because the algorithm spears ahead to find the farthest nodes and comes back to the first set of adjacent nodes at the end. The final sequence of visited nodes consists of the nodes that are always at the end of the stack list during each iteration.

10.2 Single source shortest path

We now turn to some of the most useful algorithms in network analysis, the shortest path algorithms. There are many algorithms that can be used to return the shortest paths on networks. Some of them are specifically designed to solve the problem when a single source is specified and we need to know the shortest path from that source to a target node on the network. Some algorithms are generic, while others can only deal with edges without negative weights. There are also algorithms that return the shortest path between any pair of nodes. In this section, we introduce Dijkstra's shortest path algorithm for single source problems. In the next section we will discuss an algorithm for all node pairs.

Figure 10.2 illustrates the steps involved in running Dijkstra's algorithm on our simple network, where node 1 is specified as the source. This algorithm does not need to specify the target node because its result will allow us to know the shortest paths (and the

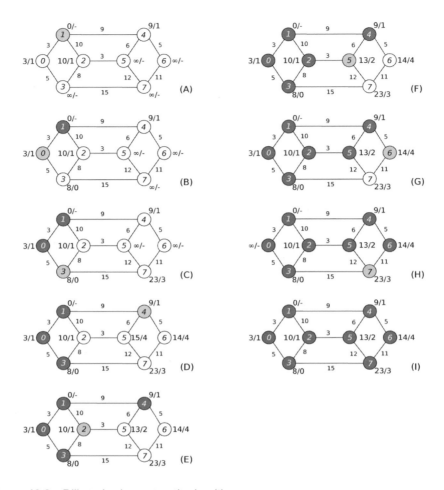

Figure 10.2 Dijkstra's shortest path algorithm

cumulative weights or distances) from the source to all other nodes. During each step of the algorithm, we will need to assign two values to each node: the cumulative distance from the source by the shortest path and the immediately previous node, and we show them as two symbols or numbers separated by a slash. When the algorithm starts, we work on the source node and we mark it in light grey. At this point, we only know these values for the source, which gets 0/–, indicating the cumulative weight is zero and there is no immediately previous node. The adjacent nodes of the source will be assigned a weight using their weights from the source and the previous node as the source node. At this initial stage we assign all other nodes the values ∞/–, meaning that we have not computed the cumulative distance and do not yet know the previous node on the path (Figure 10.2A).

We are done with node 1 and mark it in dark grey. Next, we examine the adjacent nodes. These are nodes 0, 2, and 4 in our case. The next node we will need to work on is the node with the smallest cumulative weight at this step, excluding the dark grey nodes that we will not make any changes to anymore because we know exactly their cumulative weights and previous nodes. The next node now is node 0, which has the smallest cumulative weight of 3 among all the nodes (except node 1). We mark node 0 in light grey. We then update the values of all the nodes adjacent to node 0. We only update those if the sum of the cumulative weight of node 0 and the weight on the edge from 0 to the adjacent node is smaller than the current cumulative weight on that adjacent node. In that case, the adjacent node gets a new cumulative weight and previous node that is the node we are working on (i.e., 0). Figure 10.2 B reflects the result of this step where the cumulative weight on node 3 changes from ∞ to 8. It should be clear that we do not update the nodes that are already marked in dark grey.

Node 0 is thus done and we mark it in dark grey and continue to find the next node (i.e., the one with the smallest cumulative weight) to work on, marking it in light grey, and update its adjacent nodes (Figure 10.2C). We repeat this process until all nodes are marked in dark grey (Figure 10.2I). At this point, deriving the shortest path from the source to any node is a simple process of moving from the target node backward, guided by the previous node marked on each node. The target node should also be assigned a cumulative weight that tells the weight from the source following the shortest path. For example, the shortest path from node 1 to node 6 will be 1–4–6, with a total weight of 14.

The Python program in Listing 10.5 implements Dijkstra's algorithm. In the code, we dynamically maintain three lists: the cumulative distance for each node (`dist`), the previous node for each node (`prev`), and the candidate nodes that are not marked in dark grey (`Q`). We initially set all the distances to `INF` except for the source node (line 17). Then we repeat until the list of candidate nodes is empty (line 20). During each iteration, we find the candidate node with the smallest distance and remove it from list `Q` (line 21). Then we find the neighbors of the candidate node (line 22). For each adjacent node, we check if the update condition is met (line 25) before an update is done. We use two functions to find the candidate node and remove it from `Q` (`get_remove_min`) and to find the neighbors of any node (`get_neighbor`). Finally, the `shortest_path` function uses the result from Dijkstra's algorithm and assembles the actual shortest path between the source and any given target node.

Listing 10.5: Dijkstra's single source shortest path algorithm (dijkstra.py).

```
from network2listmatrix import *

def dijkstra(source, distmatrix):
    """
    Dijkstra's single source shortest path algorithm.
    Input
      source: index to source node
      distmatrix: distance matrix where cell (i, j) is INF
                  if nodes i and j are not on an edge,
                  otherwise the value is the weight/distance
                  of the edge
    Output
```

```
           dist: cumulative distance from source to each node
           prev: list of previous node for each node on the path
       """
       n = len(distmatrix)
       dist = [INF if i!=source else 0 for i in range(n)]
       prev = [None for i in range(n)]
       Q = range(n)
       while len(Q)>0:
           u = get_remove_min(Q, dist)
           U = get_neighbor(u, distmatrix, n)
           for v in U:
               newd = dist[u] + distmatrix[u][v]
               if newd < dist[v]:
                   dist[v] = newd
                   prev[v] = u
       return dist, prev

def get_remove_min(Q, dist):
    """
    Finds the node in Q with smallest distance in dist, and
    removes the node from Q.
    Input
      Q: a list of candidate nodes
      dist: a list of distances to each node from the source
    Output
      imin: index to the node with smallest distance
    """
    dmin = INF
    imin = -1
    for i in Q:
        if dist[i] < dmin:
            dmin = dist[i]
            imin = i
    Q.remove(imin)
    return imin

def get_neighbor(u, d, n):
    neighbors = [i for i in range(n)
                 if d[i][u]!=INF and i!=u]
    return neighbors

def shortest_path(source, destination, distmatrix):
    dist, prev = dijkstra(source, distmatrix)
    last = prev[destination]
    path = [destination]
    while last is not None:
        path.append(last)
        last = prev[last]
    return path, dist[destination]
```

```
63  if __name__ == "__main__":
64      fname = '../data/network-links'
65      a = network2distancematrix(fname, True)
66      print shortest_path(1, 6, a)
67      print shortest_path(0, 7, a)
```

Here we use the `network2distancematrix` function to read the network file into a matrix form that is then used by the `shortest_path` and `dijkstra` functions. Running the test part of the code yields the following result:

```
([6, 4, 1], 14)
([7, 3, 0], 20)
```

The shortest paths between nodes 6 and 1 and between nodes 7 and 0 can be confirmed by examining Figure 10.2. The printout of the path can be easily reversed so that the list starts from the source.

Shortest path algorithms like Dijkstra's are no doubt very useful, and indeed are in widespread and constant use. For example, online routing services are commonly used to search for driving and other navigation purposes. Here we introduce another special use of the shortest path algorithm in a context called the gateway problem. We wish to find the shortest path between two locations, but we require the path to go through a particular node, the gateway, for some reason. In landscape planning, for example, we often want a path to go through a scenic location. We can actually achieve this by running Dijkstra's algorithm twice, once from each of the two endpoints, and then we can easily construct the path through any given gateway. The code in Listing 10.6 implements this idea.

Listing 10.6: Using Dijkstra's algorithm to solve the gateway problem (gateway.py).

```
 1  from network2listmatrix import *
 2  from dijkstra import shortest_path
 3  
 4  def gateway(s1, s2, gatewaynode, distmatrix):
 5      """
 6      Finds the shortest path through a gateway location.
 7      Input
 8        s1: one end node
 9        s2: another end node of the path
10        gatewaynode: node for the gateway location
11      Output
12        A list of the nodes on the path, and the total weight
13      """
14      path1, d1 = shortest_path(s1, gatewaynode, distmatrix)
15      path2, d2 = shortest_path(s2, gatewaynode, distmatrix)
16      path1.reverse()
17      return path1[:-1]+path2, d1+d2
18  
19  if __name__ == "__main__":
20      fname = '../data/network-links'
21      a = network2distancematrix(fname, True)
22      print gateway(0, 7, 2, a)
```

10.3 All pair shortest paths

While Dijkstra's algorithm deals efficiently with shortest path problems from a single source, if we use it to find the shortest paths among all pairs of nodes in a network, we have to run the algorithm for each node as the source. This turns out to be inefficient and we need a new algorithm for that purpose. Here we introduce the Floyd–Warshall algorithm for all node pairs.

This is an iterative process, and we use the index k to refer to the iteration number, with $k = 0$ for the initial step. We maintain a set of distance matrices D_k for the kth iteration, and each cell in the matrix is denoted by d_{ij}^k. For $k = 0$, we have d_{ij}^k set to infinity if nodes i and j are not directly connected by an edge. The algorithm has n iterations, where n is the number of nodes. For each iteration k, we update the matrix using the equation

$$d_{ij}^k = \min\{d_{ij}^{k-1}, d_{ik}^{k-1} + d_{kj}^{k-1}\},$$

where min is an operation that returns the smaller value in the input list of values. After n iterations, each cell $d_{ij}^n \in D_n$ stores the cumulative weight on the shortest path between nodes i and j.

It is straightforward to implement the Floyd–Warshall algorithm as long as we can maintain the distance matrix during the process of iterations, as shown in the Python code in Listing 10.7. The actual computation does not need to save the intermediate matrices. Instead we only need one matrix throughout the entire process.

Listing 10.7: The Floyd–Warshall algorithm (allpairdist.py).

```
from network2listmatrix import network2distancematrix

def allpairs(a):
    """
    Returns the weight/distance matrix for all pair
    shortest path using the Floyd-Warshall algorithm.
    Input
      a: initial distance matrix where weights for
         non-adjacent pairs are infinity
    Output
      The function directly changes the values in the input
    """
    n = len(a)
    for k in range(n):
        for i in range(n):
            for j in range(n):
                if a[i][j] > a[i][k]+a[k][j]:
                    a[i][j] = a[i][k]+a[k][j]

if __name__ == "__main__":
    fname = '../data/network-links'
    a = network2distancematrix(fname, True)
    allpairs(a)
    print a[1][6]
    print a[0][7]
```

The output of the program gives exactly the same distances between nodes 1 and 6 and between 0 and 7 as did Dijkstra's algorithm earlier (i.e., 14 and 20). Table 10.1 shows how the distance matrix changes through the iterations, where $k=0$ indicates the initial distance matrix.

An obvious problem in the all pair shortest path algorithm is the lack of path information: we know the total weight on each shortest path, but we do not know the actual path. The paths, however, can be easily recovered. Recall that we update the distance matrix as $d_{ij}^k = d_{ik}^{k-1} + d_{kj}^{k-1}$. This means that there is a shorter path from i to j by going through k. Therefore, node k must be on the path and we just need a data structure to record that, updating it whenever we update the distance between i and j again. Now we add a new matrix to store the path. It is possible to use this new matrix to store either the previous nodes on a path (as we did in Dijkstra's algorithm) or the next nodes on the path. We demonstrate the use of the next nodes here in Listing 10.8 (see the exercises for the other method that uses the previous nodes). More specifically, we maintain a matrix $\{b_{ij}\}$ such that for the shortest path from i to j the node immediately after i is b_{ij}. In our implementation, we name this matrix `next`. The matrix must be initialized so that $b_{ij} = i$ if there is an edge between i and j, otherwise we set it to -1 (line 20). Given the completed `next` matrix, the function `recoverpath` recovers the path between from node `s` to `t` by continuously replacing the start node `s` with the one indicated by `next[s][t]` (line 30) until the new `s` equals `t` (line 29).

Listing 10.8: The Floyd–Warshall algorithm with path recovery (allpairpaths.py).

```
from network2listmatrix import network2distancematrix, INF

def allpairswpaths(a, next):
    """
    Returns the weight/distance matrix for all pair shortest
    path, along with the actual path, using the
    Floyd-Warshall algorithm.
    Input
      a: initial distance matrix where weights for
         non-adjacent pairs are infinity
    Output
       The function directly changes the values in the input
    """
    n = len(a)
    for k in range(n):
        for i in range(n):
            for j in range(n):
                if a[i][j] > a[i][k]+a[k][j]:
                    a[i][j] = a[i][k]+a[k][j]
                    next[i][j] = next[i][k]

def recoverpath(s, t, next):
    """
```

```
24          Returns the paths from node u to v using the next matrix.
25          """
26          if next[s][t] == -1:
27              return None
28          path = [s]
29          while s != t:
30              s = next[s][t]
31              path.append(s)
32          return path
33
34    if __name__ == "__main__":
35        fname = '../data/network-links'
36        a = network2distancematrix(fname, True)
37        n = len(a)
38        next=[[i if a[i][j] != INF else -1
39               for i in range(n)]
40               for j in range(n)]
41        allpairswpaths(a, next)
42        print recoverpath(1, 6, next), a[1][6]
43        print recoverpath(0, 7, next), a[0][7]
```

Running the test part of the code in the above algorithm returns the following result:

```
[1, 4, 6] 14
[0, 3, 7] 20
```

These are exactly the same values as we obtained using Dijkstra's algorithm twice for the two pairs of nodes (note the order of nodes is reversed here). We note that the sequences from the two methods may not be exactly the same all the time. This is because there may be multiple shortest paths with the same total weight. For example, the path from node 6 to 3 may be either 6–4–5–2–3 or 6–4–1–0–3, both with the same weight of 22. The two algorithms have different mechanisms to break ties and thus cause different shortest paths to be taken during the iterations. In our implementation of Dijkstra's algorithm, we break ties by always choosing whichever is the first index that has the smallest cumulative distance. In the Floyd–Warshall algorithm, however, once the shortest path is found (in Table 10.1, this occurs at iteration 5), we will not take any other alternative shortest paths.

Table 10.1 Distance matrix at each iteration of the Floyd–Warshall algorithm. The number in the upper left-hand corner of each matrix is the k value.

0	0	1	2	3	4	5	6	7
0	0	3	–	5	–	–	–	–
1	3	0	10	–	9	–	–	–
2	–	10	0	8	–	3	–	–
3	5	–	8	0	–	–	–	15
4	–	9	–	–	0	6	5	–
5	–	–	3	–	6	0	–	12
6	–	–	–	–	5	–	0	11
7	–	–	–	15	–	12	11	0

1	0	1	2	3	4	5	6	7
0	0	3	–	5	–	–	–	–
1	3	0	10	8	9	–	–	–
2	–	10	0	8	–	3	–	–
3	5	8	8	0	–	–	–	15
4	–	9	–	–	0	6	5	–
5	–	–	3	–	6	0	–	12
6	–	–	–	–	5	–	0	11
7	–	–	–	15	–	12	11	0

2	0	1	2	3	4	5	6	7
0	0	3	13	5	12	–	–	–
1	3	0	10	8	9	–	–	–
2	13	10	0	8	19	3	–	–
3	5	8	8	0	17	–	–	15
4	12	9	19	17	0	6	5	–
5	–	–	3	–	6	0	–	12
6	–	–	–	–	5	–	0	11
7	–	–	–	15	–	12	11	0

3	0	1	2	3	4	5	6	7
0	0	3	13	5	12	16	–	–
1	3	0	10	8	9	13	–	–
2	13	10	0	8	19	3	–	–
3	5	8	8	0	17	11	–	15
4	12	9	19	17	0	6	5	–
5	16	13	3	11	6	0	–	12
6	–	–	–	–	5	–	0	11
7	–	–	–	15	–	12	11	0

4	0	1	2	3	4	5	6	7
0	0	3	13	5	12	16	–	20
1	3	0	10	8	9	13	–	23
2	13	10	0	8	19	3	–	23
3	5	8	8	0	17	11	–	15
4	12	9	19	17	0	6	5	32
5	16	13	3	11	6	0	–	12
6	–	–	–	–	5	–	0	11
7	20	23	23	15	32	12	11	0

5	0	1	2	3	4	5	6	7
0	0	3	13	5	12	16	17	20
1	3	0	10	8	9	13	14	23
2	13	10	0	8	19	3	24	23
3	5	8	8	0	17	11	22	15
4	12	9	19	17	0	6	5	32
5	16	13	3	11	6	0	11	12
6	17	14	24	22	5	11	0	11
7	20	23	23	15	32	12	11	0

7	0	1	2	3	4	5	6	7
0	0	3	13	5	12	16	17	20
1	3	0	10	8	9	13	14	23
2	13	10	0	8	9	3	14	15
3	5	8	8	0	17	11	22	15
4	12	9	9	17	0	6	5	16
5	16	13	3	11	6	0	11	12
6	17	14	14	22	5	11	0	11
7	20	23	15	15	16	12	11	0

6	0	1	2	3	4	5	6	7
0	0	3	13	5	12	16	17	20
1	3	0	10	8	9	13	14	23
2	13	10	0	8	9	3	14	15
3	5	8	8	0	17	11	22	15
4	12	9	9	17	0	6	5	18
5	16	13	3	11	6	0	11	12
6	17	14	14	22	5	11	0	11
7	20	23	15	15	18	12	11	0

10.4 Notes

All the algorithms introduced in this chapter can be found in different texts, especially the book by Cormen et al. (2001) that provides a thorough account of these algorithms. The running times for the algorithms discussed in this chapter are typically determined by the number of nodes (N) and edges (E) of the network. Both the breadth-first and depth-first search algorithms have a running time of $O(N+E)$. The running time of Dijkstra's algorithm (Dijkstra, 1959) is likewise determined by the number of nodes (N) and edges (E). However, depending on how the algorithm is implemented, the actual (still theoretical) running time may be different. In our implementation, where, since we use a linear search to find the minimal value in Q (in the function get_remove_min), the computing time is $O(N^2)$. However, it is possible to speed this up using a different way of searching for the minimal value and achieve a running time of $O(N + E \log E)$. The three loops in the Floyd–Warshall algorithm (Floyd, 1962) clearly make the time complexity of this algorithm $O(N^3)$, which is theoretically more efficient than running

Dijkstra's algorithms for all pairs. The gateway problem (Lombard and Church, 1993) is an interesting use of Dijkstra's algorithm, and more applications of the shortest path algorithms can be found elsewhere, especially in the transportation literature (Miller and Shaw, 2001).

10.5 Exercises

1. Design your own network using the one in Figure 10.1 as an example. Write the network input file for your network and run the algorithms discussed in this book.

2. Both the functions `network2list` and `network2distancematrix` have an input argument called `is_zero_based`. This can be used to specify if the nodes in the network input file are indexed on a zero basis. With a small change to the program you can actually allow any base for the indices, of course as long as they are consistent in a file. In the next chapter we will introduce a lot of networks of different sizes. Try to run the algorithms on those files and test the complexity of these algorithms.

3. Write a Python program to generate random networks. First create a number of nodes, and then start to randomly insert edges between the nodes that have not been linked. As you increase the number of edges, the properties of the network will change. For example, the average shortest path will decrease from infinity (when the network is not connected). By just using this simple characteristic of networks you can examine the connectivity of the network. For each network setting, defined by the number of nodes and random edges, you will need to do this many times because each time we will have a different network. Plot the relationship between the number of random edges and the connectivity of the network.

4. Can we use the depth-first or breadth-first search algorithm to find the shortest path between two nodes? Why or why not?

5. We talked about the theoretical difference if Dijkstra's algorithm is actually used to retrieve all the pair shortest paths. Can you empirically test the theory?

6. The Python documentation has argued that using a list as a queue is not efficient. Instead, we could use a module called `deque`. Replace the implementation of queues in the breadth-first search with `deque`. Can you actually tell the difference?

7. Rewrite the code for the Dijkstra's algorithm using the list representation of networks.

8. In the all pair Floyd–Warshall algorithm with path recovery, we use a new matrix called `next` to store the *next* nodes on the path. By a similar logic, we can store the *previous* nodes in a patch starting from the target node. Modify the code to implement this idea.

9. We have argued in the main text that both the Dijkstra and Floyd–Warshall algorithms will return the same result in terms of distance but may return different shortest paths. Write a Python program to test this. Can you also identify why they sometimes report different paths? Is it possible to force both algorithms to return the same paths?

10. Write a Python program to report the number of edges for each node. This is one of the many ways we can measure a network, and this specific number is called the nodal degree, indicating the involvement of each node in the network.

11. How can we compute the average shortest distance between node pairs in a network? This value indicates the closeness of nodes in the network. In the social networks literature, this is also called closeness centrality.

12. Write a Python program that can be used to generate networks with high closeness centrality.

13. A node in a network may be important because many paths between other nodes go through it. Write Python code to find which node has the highest number of shortest paths going through it. In the social networks literature this is called the betweenness centrality.

14. Write a Python program that can be used to generate networks with high betweenness centrality.

Spatial Optimization

Agent Smith: We'll need a search running.
Agent Jones: It has already begun.

The Matrix (1999)

Many problems in the world demand solutions that are deemed to be the best or optimal, and we believe these optimal solutions can be found by searching through many alternative solutions to the problem. For any such problem, we have a way to evaluate how an alternative solution performs so that we can tell one solution is "better" than another. We use an objective function to tell how a solution can be evaluated and compared with others. Of course being better may be subjective, but when it comes to solving an optimization problem we will need a formal objective function so that all alternative solutions can be measured and compared. In this chapter, we focus on the optimization problems that contain some spatial components that must be considered in determining the optimal solution. In many applications, our aim is often to find a set of locations that are measured to be the best when compared with other solutions. Take the problem of locating a cellular tower for a wireless phone company, for example. The aim here is to find a place where a tower can be located so that it covers as many customers as possible. There will be many locations where the tower can cover many customers. But we believe there is at least one location where we can cover more customers than at any other location. This is the optimal location and hence the optimal solution to the problem. A similar example comes from forest fire monitoring, where watchtowers must be located as part of an alarm system. When the concept of being covered has a different meaning, we have a different problem. For example, when we plan to find a place for a public library, we do not say that this library only serves a certain portion of the general public. Instead, all are served. What would be a good indication of such a library being optimal? We can make the case for many different measures. From a spatial perspective, we can say that a location is efficient if it is convenient for the library patrons to visit. Convenience can be measured by the average time patrons have to spend getting to the location. Therefore, among all possible locations, there is at least one location that has the shortest average time (or distance, treated as equivalent to time).

There are many kinds of spatial optimization problem. In this chapter we discuss a few of these, and our main focus is on the algorithms that can be used to solve these problems. In general, two kinds of solution method can be used. The first is the exact methods that, as their name suggests, can be used to find exactly the optimal solutions. There are many ways to find the exact solution. For example, the shortest path algorithm introduced in Chapter 10 can find exactly the shortest path in a network, which is achieved by dynamically adjusting the shortest path found so far in each iteration of the process. We can also think of a brute-force approach in which we exhaustively search all possible solutions and find the best one. However, an optimization problem typically has a huge number of possible solutions, which together are referred to as the solution space of the problem. The actual size of a solution space depends on the actual problem to be solved. The number of variables in a problem is the main factor that determines the size of the solution space, which often increases quickly with the number of variables. There are many optimization problems that share such a common feature: they are difficult to solve because their solution space can be huge. For example, in the above public library example, suppose we have 100 potential locations from which we need to choose five in which to put new libraries. The total number of possible choices is a number called "100 choose 5," which equals $100! / (5! \times 95!) = 75,287,520$. While this is a large number, let us think about another example. Suppose we have 20 cities; we would like to find a tour that visits each city once and only once, in such a way as to minimize the total distance (or cost) of the tour. We simply want a sequence that indicates the order of cities to be visited, and we know the distance between every pair cities. This will be a sequence of 20 cities. At the ith ($0 \leq i \leq 19$) position in the sequence, we have $20 - i$ choices, and in total we have $20! = 2,432,902,008,176,640,000$ choices. Huge, for only 20 cities. This is a very difficult problem called the traveling salesman problem (TSP).

It should be clear at this point that searching through the solution space in order to find the optimal solution can be *very* time-consuming. To take the TSP with 20 cities as an example, if a computer can evaluate 1 billion tours per second, it will take 2,432,902,008 seconds to exhaustively evaluate each tour, which is more than 77 years. But how realistic is a speed of 1 billion tours per second? We will test this using the following simple Python program:

```
import time

time1 = time.time()
sum = 0
while sum < 1000000000:
    sum += 1
time2 = time.time()
print sum, time2-time1
```

The output is not optimistic: it took 122.10 seconds on my MacBook Air (1.3 GHz Intel Core i5). This is awfully long! But we know Python is an interpreter that does not compile the program. We now run a compiled one-line C program[1] on the same computer. This time it is better: it took slightly more than 4 seconds. Here we did nothing more than simple looping 1 billion times. An actual evaluation of a tour will need to get all the

---

[1] `int main() {double sum = 0; while (sum < 1000000000) sum += 1; }`

distances and compute the sum. This means it would take a really good (fast) computer to achieve this speed.

It is obvious, then, that we will need to do something smart in order to find the best solution out of so many. Otherwise, searching for and evaluating each and every one of the solutions will quickly become impractical when the problem gets large (more cities in the TSP, for example). The shortest path algorithm we mentioned before is such an example because it does not examine each possible path to get the optimal one. It ignores a large portion of the possible solutions to achieve a very decent computing time. This is also how some of the exact algorithms were designed: they try to utilize the structure of the problem to reduce the size of solution space.

Unfortunately, not all exact methods can effectively reduce the size of the solution space. Sometimes, though, we may not be so interested in a solution being optimal and we will be quite satisfied when a suboptimal solution is found. There is a more fundamental reason for accepting suboptimal solutions: a solution is only optimal in the context of how the problem is formulated. When it comes to reality, the optimal solution in theory may become much worse because several factors are not considered in the original problem. Of course, we also want to find good solutions quickly. This is where heuristic methods come in: the methods that cannot guarantee to find the optimal solution, but can be used to find high-quality, near-optimal solutions quickly.

We will discuss these two categories of solution approaches in turn. We discuss some exact methods in the context of location problems in this chapter, and then we move our focus in the next chapter to various heuristic methods. Exact algorithms come in different forms. The shortest path algorithms belong to a category of exact methods called dynamic programming. It is impossible to exhaust all kinds of optimization problem or their solution methods, nor is that the purpose of this chapter. We focus on the way these algorithms are designed and how they are implemented using our knowledge about geographic data. In this chapter, we explore two more kinds of exact search algorithm. The first concentrates on the problem of finding the center to a set of points, and the algorithms developed to solve this problem all utilize the geometric properties of such a problem in some way. The second type represents a more generalized approach: we try to reduce the optimization problem at hand to a general problem and then the solution methods for the generalized problem can be used to solve our own problem. The general problem here is linear programming, and a lot of optimization problems can be reduced to a linear programming problem. We explore how such an approach can be used in another useful spatial optimization problem called the p-median problem.

11.1 1-center location problem

Given a set of points, what is the smallest circle that encloses all the points? This is a problem that has been studied under many different names, including 1-center location problem, minimal enclosing circle, and smallest enclosing disc problem. From the perspective of location-allocation modeling, we tend to call it the 1-center location problem, but this is a generic problem that has a wide range of applications in different domains. Here, we are given a finite set of points in space. The circle can be determined given its center and radius, and it is obvious that the center can be at any location on the continuous space. We will introduce three algorithms that can be used to solve this problem exactly, and we will test these algorithms in terms of their efficiency. All the algorithms discussed here start with an initial circle and then find their way to the optimal circle so that all the points are enclosed. The initial circle can be a small circle that only encloses a subset of the points, or a large circle that already encloses all the points. Then an iterative process is devised to make sure the circle either enlarges or shrinks before the optimal one is reached.

Before we dive into the algorithms, however, let us sort out a few data structure issues. We need a data structure that can be used for all the algorithms. Specifically, Listing 11.1 includes a Python class called `disc` designed for that purpose. Line 3 makes sure the Python interpreter knows the path to the `Point` class that is stored in the point.py file (Listing 2.1) in the `geom` directory at the same hierarchical level as the current directory of this file.[2] The `disc` class mainly contains two critical pieces of information: the center and radius of the circle. A `disc` instance can be initialized in two ways: using the center point and radius explicitly (line 10) or a set of two or three points (line 13). In the latter case, we will need to compute the center and radius of the circle that passes through the two or three points specified, which is done using one of the functions included in the code. These are fundamental geometric computations and we do not go into the details in this book; they will be used in other programs in this chapter and the next.

Listing 11.1: Data structure used for the algorithms for the 1-center location problem (disc.py).

```
1   from math import fabs, sqrt
2   import sys
3   sys.path.append('../geom')
4   from point import *
5
6   __all__ = ['disc']
7
8   class disc:
9       def __init__(self, center=None, radius=None, points=None):
10          if points == None:
11              self.center = center
12              self.radius = radius
13          else:
14              if len(points)==2:
15                  res = make_disc2(points[0],points[1])
16              elif len(points)==3:
17                  res = make_disc(points[0],points[1],points[2])
18              else:
19                  res = [None, None]
20              self.center = res[0]
21              self.radius = res[1]
```

---

[2] Note that all the Python programs are stored in different subdirectories under the `programs` directory. This is easy for managing the code for the same topic, but it will require us to append the path to other classes at the beginning of the program.

```
22      def __eq__(self, other):
23          return self.center==other.center and\
24              self.radius==other.radius
25      def __repr__(self):
26              return "({0}, {1})".format(
27                  self.center, self.radius)
28      def inside(self, p):
29          dx = fabs(self.center.x - p.x)
30          dy = fabs(self.center.y - p.y)
31          if dx>self.radius or dy>self.radius:
32              return False
33          if self.center.distance(p) <= self.radius:
34              return True
35          return False
36
37  def make_disc2(p1, p2):
38      dx = fabs(p1.x - p2.x)
39      dy = fabs(p1.y - p2.y)
40      radius = sqrt(dx*dx + dy*dy)/2.0
41      x = min(p1.x, p2.x) + dx/2.0
42      y = min(p1.y, p2.y) + dy/2.0
43      return Point(x, y), radius
44
45  def make_disc(p1, p2, p3):
46      x1, x2, x3 = p1.x, p2.x, p3.x
47      y1, y2, y3 = p1.y, p2.y, p3.y
48      a = fabs(x2-x1)
49      b = fabs(x3-x1)
50      c = fabs((y2-y1)/a - (y3-y1)/b)
51      xs = a
52      if b < xs:
53          xs = b
54      if c < xs:
55          xs = c
56      a = fabs(y2-y1)
57      b = fabs(y3-y1)
58      c = fabs((x2-x1)/a - (x3-x1)/b)
59      ys = a
60      if b < ys:
61          ys = b
62      if c < ys:
63          ys = c
64      if xs < ys:    # eliminate x, compute y first
65          return make_disc_x(p1, p2, p3)
66      else:
67          return make_disc_y(p1, p2, p3)
68
69  def make_disc_y(p1, p2, p3):
70      x1 = p1.x
71      x2 = p2.x
72      x3 = p3.x
```

```
73        y1 = p1.y
74        y2 = p2.y
75        y3 = p3.y
76        t1 = (x1*x1-x3*x3+y1*y1-y3*y3)/(2*(x3-x1))
77        t2 = (x1*x1-x2*x2+y1*y1-y2*y2)/(2*(x2-x1))
78        t3 = (y2-y1)/(x2-x1) - (y3-y1)/(x3-x1)
79        y = (t1 - t2)/t3
80        x = -(2*(y2-y1)*y + x1*x1 - x2*x2 + y1*y1 -
81                y2*y2) / (2*(x2-x1))
82        r = sqrt((x1-x)*(x1-x) + (y1-y)*(y1-y))
83        return Point(x, y), r
84
85  def make_disc_x(p1, p2, p3):
86        x1, x2, x3 = p1.x, p2.x, p3.x
87        y1, y2, y3 = p1.y, p2.y, p3.y
88        t1 = (x1*x1-x3*x3+y1*y1-y3*y3)/(2*(y3-y1))
89        t2 = (x1*x1-x2*x2+y1*y1-y2*y2)/(2*(y2-y1))
90        t3 = (x2-x1)/(y2-y1) - (x3-x1)/(y3-y1)
91        x = (t1 - t2)/t3
92        y = -(2*(x2-x1)*x + x1*x1 - x2*x2 + y1*y1 -
93                y2*y2) / (2*(y2-y1))
94        r = sqrt((x1-x)*(x1-x) + (y1-y)*(y1-y))
95        return Point(x, y), r
```

The first algorithm for the 1-center problem is called the Chrystal–Peirce algorithm and has a history of more than a century (see Notes). Let us suppose that all the points are stored in a list called P. This algorithm starts with a very large circle that also passes through two points in P and encloses all the points. Without loss of generality, we call these two points A and B. We now find a third point C in P (excluding A and B) such that the angle ACB is minimized. In two conditions we have found the optimal solution. First, if the triangle ABC is non-obtuse, meaning with largest angle no more than 90 degrees, then the three points define the minimal enclosing circle. Second, if angle ACB is obtuse, the circle that has points A and B on the diameter is the minimal enclosing circle. For all other cases, we will need to continue the search process. We will call the two points on the non-obtuse angles of the triangle A and B and repeat the algorithm to search for a new point C until one of the two conditions brings us to a stop.

The original algorithm starts with a very large circle covering all the points and passing through two adjacent points on the convex hull of the set of points. However, it is time-consuming to find the convex hull and the circle that covers all the points, with a time complexity of $O(n \log n)$ for n points. A later algorithm makes finding the initial large circle more efficient, with a time complexity of $O(n)$. This new start strategy, which we call the Chakraborty–Chaudhuri method, starts by finding a point a in P that is farthest from the origin point. Then it finds another point that maximizes

$$\frac{d_a d(ab)^2}{|x_a(x_a - x_b) + y_a(y_a - y_b)|},$$

where d_a is the distance from point a to the origin, and $d(ab)$ is the distance between the two points. The circle that passes through a and b is big enough to enclose all the points in P.

Listing 11.2 shows the Chrystal–Peirce algorithm. There are four functions that provide support for various tasks. The function `get_angle` (line 8) calculates the angle between the last two points and the first point. The function `find_mini_angle` (line 24) finds the point (C) that forms the minimal angle with A and B. The function `getfirsttwo` (line 42) gets the two points that can be used to create a very large circle for the initial step of the algorithm using the Chakraborty–Chaudhuri method. The function `moveminimaxpoints` (line 64) makes sure that the two initial points are the first two items in the list.

The function `oncenter1` is the core search algorithm (line 72). We use the first two points in P as points A and B, which means that the list will need to be prepared before it can be used in this algorithm. When searching for point C in P, we do not physically remove the points that have been considered. Instead we use a list of integers called `unmarked` to indicate whether a point has been searched (line 74). Lines 76 and 77 assign points A and B. The two conditions where an optimal solution is found are checked at lines 82 and 86. For the other cases (line 89), we will use point C to replace the current A or B, depending on which point sits on an obtuse angle.

Listing 11.2: Chrystal–Peirce algorithm for the 1-center location problem (cp.py).

```
1   from math import sqrt, atan2, fabs, pi
2   from disc import *
3   import sys
4   sys.path.append('../geom')
5   from point import *
6   import random
7
8   def get_angle(p0, p1, p2):
9       """
10          Returns angle between line p0p1 and p0p2 (using
11          directions: p0 -> p1, p0 -> p2)
12      """
13      a1 = atan2(p1.y-p0.y, p1.x-p0.x)
14      a2 = atan2(p2.y-p0.y, p2.x-p0.x)
15      if a1 < 0:
16          a1 = 2*pi - fabs(a1)
17      if a2 < 0:
18          a2 = 2*pi - fabs(a2)
19      degree = fabs(a1-a2)
20      if degree > pi:
21          degree = 2*pi - degree;
22      return degree;
23
24  def find_mini_angle(S1, A, B, um):
25      n = len(S1)
26      angle = 100
27      c = -1
```

```
        for i in xrange(n):
            if S1[i] == A or S1[i] == B:
                continue
            if not um[i]:
                continue
            angle1 = get_angle(S1[i], A, B)
            if angle1 < angle:
                angle = angle1
                c = i #D = S1[i]
            if angle1 < pi/2.0:
                continue
            um[i] = 0
        return angle, c

def getfirsttwo(S):
    b = Point(0, 0)
    dist = 0
    ix = 0
    for i, p in enumerate(S): # farthest point from origin
        d1 = p.distance(b)
        if d1 > dist:
            dist = d1
            ix = i
            a = p
    dist = 0
    dd = sqrt(a.x*a.x+a.y*a.y)
    for i, p in enumerate(S):
        if i==ix: continue
        d1 = (a.x-p.x)*(a.x-p.x) + (a.y-p.y)*(a.y-p.y)
        d1 = d1*dd / fabs(a.x*(a.x-p.x) + a.y*(a.y-p.y))
        if d1 > dist:
            dist = d1
            b = p
            iy = i
    return ix, iy

def moveminimaxpoints(points):
    mini, maxi = getfirsttwo(points)
    # set minx to 0, maxx to 1
    if maxi != 0:
        points[0],points[mini] = points[mini], points[0]
    if mini != 1:
        points[1],points[maxi] = points[maxi], points[1]

def onecenter1(P):
    n = len(P)
    unmarked = [1 for i in range(n)]
    done = False
    a = 0
    b = 1
```

```
78        while not done:
79            A, B = P[a], P[b]
80            angle, c = find_mini_angle(P, A, B, unmarked)
81            C = P[c]
82            if angle > pi/2.0:
83                done = True
84                d = disc(points=[A, B])
85            elif get_angle(B, A, C) < pi/2.0 and\
86                 get_angle(A, B, C) < pi/2.0:
87                done = True
88                d = disc(points=[A, B, C])
89            else:
90                if (get_angle(B, A, C)) > pi/2.0:
91                    unmarked[b] = 0
92                    b = c
93                else:
94                    unmarked[a] = 0
95                    a = c
96        return d
97
98  def test():
99      npts = 5
100     points = []
101     for i in xrange(npts):
102         p = Point(random.random(), random.random())
103         points.append(p)
104     print points
105     print onecenter1(points)
106     moveminimaxpoints(points)
107     print points
108     print onecenter1(points)
109
110 if __name__ == '__main__':
111     test()
```

We now test the Chrystal–Peirce algorithm using just five random points in the `test` function. The first experiment does not alter the points in the list, so the first two points do not guarantee a circle that encloses all points (line 105). Then the start strategy is used (line 106). After running this many times, we can see that for the five points, the algorithm finds the best solution without using the start strategy. However, occasionally, we might get a result like this:

```
[(0.4, 0.8), (0.1, 0.3), (0.0, 0.8), (0.8, 0.4), (0.5, 0.4)]
((0.4, 0.4), 0.371971373225)
[(0.4, 0.8), (0.8, 0.4), (0.0, 0.8), (0.1, 0.3), (0.5, 0.4)]
((0.4, 0.6), 0.450773182422)
```

where it is clear the two output lines are not consistent. The algorithm does not check whether all the points are enclosed and therefore it is critical that we start with a large circle so that during the process we maintain the complete enclosing of points.

Spatial Optimization

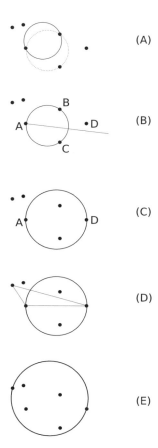

Figure 11.1 Elzinga–Hearn algorithm for the 1-center problem. (A) The initial two points and the circle (solid line). Since the initial circle does not cover all points, a new point is added and a strict acute triangle is formed. The circle of the three points (dotted) still does not cover all points. (B) An outside point (D) is chosen and the three points on the circle are marked as A, B, and C, respectively. Point C is on an obtuse angle and is therefore dropped. (C) The remaining two points (A and D) lie on a circle that does not cover all points. (D) A new point is added, but the three points are on an obtuse triangle and the one on the largest angle is dropped. (E) The final circle going through two points covers all points

From the above discussion, it is clear that a minimal enclosing circle of a set of points is determined either by two points to form the diameter or by three points that form a strict acute triangle. The Elzinga–Hearn algorithm has also been developed to utilize this fact. This algorithm starts with a circle determined by any two points in P and then goes on to continuously increase the size of the circle in three iterative steps until the optimal circle is found. (1) If the circle defined by the two points does not cover all the points, then a point outside the circle is chosen. (2) If the three points form a right or obtuse triangle, then we drop the point with the right or obtuse angle and go back to step 1 with the two remaining points as the initial points. (3) Now we have a strict acute triangle.

If the circle passing through the three points does not cover all points, we will continue our search by choosing an outside point. Let us call this outside point *D* and among the three points mark the farthest point from *D* as *A*. Then we draw a line that goes through point *A* and the center of the circle, and among the remaining two points, the one on the same side of *D* is marked as *B* and the other one as *C* (Figure 11.1B). Ties can be broken arbitrarily. Points *A*, *C*, and *D* then are used to go to step 2. Figure 11.1 illustrates the search process of this algorithm.

Listing 11.3 includes the code for the Elzinga–Hearn algorithm. We use the function `right_obtuse_triangle` to test if the triangle formed by the three input points has a right or obtuse angle. If so, it returns `True` and rearranges the list of points so the first two points in the list are associated with acute angles. The function `find_three` takes three points, the circle defined by the points, and an extra point outside the circle to find the three points *A*, *C*, and *D* that will be used for the next step. The `for` loop in line 52 finds the point *A* that is farthest from *D*. Point *D* is either above or below the line between point *A* and the center of the circle, and we can determine which by testing point *D* using the linear formula (lines 48 and 49). The `for` loop in line 52 finds the point on the same side as *D* with respect to the line between *A* and the center of the circle. In the actual computation, when points *A*, circle center, *D*, and one of the other two points are collinear or close to collinear (line 59), simply checking which side each point lies on may not work and we will use the point that has the smallest value in the line equation as point *C*. The `cover_all` function simply checks if a circle covers all points, where line 72 is necessary in a computing environment because all floating point numbers in Python are not truly real numbers and the precision of the number will cause a slight difference between the true distance between two points and the distance calculated. For this reason, we have to manually exclude those points that are used to make the circles from being tested. The function actually returns two variables: `True` or `False` if all points are covered, and the first point that is not covered, if any. The implementation of the algorithm uses the first point found here as the outside point to proceed. The actual function for the algorithm (`onecenter2`) is short and matches the description in the above paragraph (line 78).

Listing 11.3: Elzinga–Hearn algorithm (elzinga_hearn.py).

```
1   from math import pi
2   import random
3   from disc import *
4   import sys
5   sys.path.append('../geom')
6   from point import *
7   from cp import get_angle
8
9   def right_obtuse_triangle(p3):
10      """
11      if true, set p3[0] and p3[1] to define the longest edge
12      and p3[2] is on the angle >= 90
13      """
14      angle0 = get_angle(p3[0], p3[1], p3[2])
15      angle1 = get_angle(p3[1], p3[0], p3[2])
16      angle2 = pi - angle0 - angle1
```

```
17        maxa = -1.0
18        maxi = -1
19        for i, a in enumerate([angle0, angle1, angle2]):
20            if a > maxa:
21                maxa = a
22                maxi = i
23        if maxa >= pi/2.0:
24            if maxi != 2:
25                p3[maxi], p3[2] = p3[2], p3[maxi]
26            return True
27        return False
28
29    def find_three(p3, D, d):
30        """
31        Given three points in p3, an outside point D, and the
32        disc d, find A, C, D and assign them to p3[0], p3[1],
33        and p3[2], respectively
34        """
35        maxd = 0
36        for i in range(len(p3)):
37            tmpd = p3[i].distance(D)
38            if tmpd > maxd:
39            maxd = tmpd
40            iA = i
41        x1 = p3[iA].x
42        x2 = d.center.x
43        y1 = p3[iA].y
44        y2 = d.center.y
45        a = y2-y1
46        b = -(x2-x1)
47        c = (x2-x1)*y1 - (y2-y1)*x1
48        eqd = a*D.x + b*D.y + c
49        positive = eqd > 0
50        eq = [0 for i in range(3)]
51        iC = -1
52        for i in range(3):
53            if i==iA:
54                eq[i] = 0.0
55            else:
56                eq[i] = a*p3[i].x + b*p3[i].y + c
57                if positive != ((eq[i]>0)):
58                    iC = i
59        if iC == -1:
60            tempf = 100000000.0
61            for i in range(3):
62                if i == iA:
63                    continue;
64                if fabs(eq[i]) < tempf:
65                    tempf = fabs(eq[i])
66                    iC = i
67        p3[0], p3[1], p3[2] = p3[iA], p3[iC], D
68        return
```

```python
def cover_all(points, d, pp):
    for p in points:
        if p in pp: # handle precision in float numbers
            continue
        if not d.inside(p):
            return False, p
    return True, None

def onecenter2(P):
    p3 = [Point(-1, -1) for i in range(3)]
    p3[0] = P[0]
    p3[1] = P[1]
    d = disc(points=[p3[0], p3[1]])
    n = len(P)
    cnt = 0
    stop, p3[2] = cover_all(P, d, p3[:2])
    while not stop:
        if right_obtuse_triangle(p3): # right/obtuse triangle
            d = disc(points=[p3[0], p3[1]])
            stop, p3[2] = cover_all(P, d, p3[:2])
        else:                          # strict acute triangle
            d = disc(points=[p3[0], p3[1], p3[2]])
            stop, pd = cover_all(P, d, p3) # pd outside d
            if not stop:
                find_three(p3, pd, d)
        cnt += 1
    return d

def test():
    npts = 50
    points = []
    for i in xrange(npts):
        p = Point(random.random(), random.random())
        points.append(p)
    print onecenter2(points)

if __name__ == '__main__':
    test()
```

Testing the above algorithm is straightforward. Below is an example output from running the `test` function in the code:

```
[(0.1, 0.3), (0.9, 0.5), (0.6, 0.9), (0.1, 0.8), (0.6, 0.6)]
((0.5, 0.5), 0.440163316231)
```

The third algorithm for the 1-center problem is the Welzl algorithm, which starts with a small circle covering the first two points listed in the data and continuously increases the size of the circle by incorporating additional points. The code in Listing 11.4 shows how this is done. The function `minidisc` (line 27) is the core function where we first create a minimal disc that encloses the first two points in the data (line 30). It then uses a loop to keep adding additional points from the data (line 31) until all the points are considered. At each iteration of the loop, we create a new minimal disc that covers the points so far plus the additional point q. This is done in the function `minidiscwithpoint` (line 35). Now that we know the new point, q, is not in the previous disc, it must be on the boundary of a new, larger circle since the circle has to cover it. We use `minidiscwithpoint` to make sure the new circle includes q and other points covered so far on the circle boundary. In this function, we start with a circle that includes q and another point (line 18). Then we keep considering additional points. When the point is covered by the disc (line 21), we do not need to make the circle bigger, which is a way of checking circles determined by two points. Otherwise, we call the `minidiscwith2points` function to create a circle that includes three points: q, the new point, and another point from the data searched so far (line 24). Running the test function in the code should return results similar to those we obtained with the Elzinga–Hearn algorithm.

Listing 11.4: Welzl algorithm for the 1-center location problem (welzl.py).

```
import random
from disc import *
import sys
sys.path.append('../geom')
from point import *

def minidiscwith2points(P, q1, q2, D):
    D[0] = disc(points = [q1, q2])
    n = len(P)
    for k in range(n):
        if D[k].inside(P[k]):
            D[k+1] = D[k]
        else:
            D[k+1] = disc(points=[q1, q2, P[k]])
    return D[n]

def minidiscwithpoint(P, q, D):
    D[0] = disc(points = [P[0], q])
    n = len(P)
    for j in range(1, n):
        if D[j-1].inside(P[j]):
            D[j] = D[j-1]
        else:
            D[j] = minidiscwith2points(P[:j], P[j], q, D)
    return D[n-1]

def minidisc(P):
    n = len(P)
    D = [ disc() for i in range(n) ]
    D[1] = disc(points=[P[0], P[1]])
```

```
        for i in range(2, n):
            if D[i-1].inside(P[i]):
                D[i] = D[i-1]
            else:
                D[i] = minidiscwithpoint(P[:i], P[i], D)
    return D[n-1]

def test():
    n = 5
    points = []
    for i in xrange(n):
        p = Point(random.random(), random.random())
        points.append(p)
    print points
    print minidisc(points)

if __name__ == '__main__':
    test()
```

We now test the performance of the three algorithms introduced in this section. The program in Listing 11.5 sets up experiments on 50,000 points. Two data preparation scenarios are tested. First, we simply use random points as the start strategy for all algorithms. Then we test the impact of the Chakraborty–Chaudhuri algorithm by starting each search with two points in the data that are far apart.

Listing 11.5: Testing algorithms for the 1-center problem (test_1center.py).
```
from elzinga_hearn import *
from welzl import *
from cp import *

import time

n = 50000
points = [ Point(random.random(), random.random())
          for i in range(n) ]

#########################################
#
# Test performance using random start
#
#########################################

time1 = time.time()
d1 = minidisc(points)
time2 = time.time()
d1t = time2-time1

```

```
22  d2 = onecenter2(points)
23  time3 = time.time()
24  d2t = time3-time2
25
26  d3 = onecenter1(points)
27  time4 = time.time()
28  d3t = time4-time3
29
30  print "Welzl              ", d1t, d1
31  print "Elzinga-Hearn      ", d2t, d2
32  print "Chrystal-Peirce    ", d3t, d3
33
34  ##########################################
35  #
36  # Test performance after data preparation
37  #
38  ##########################################
39
40  time0 = time.time()
41  moveminimaxpoints(points)
42
43  time1 = time.time()
44  d1 = minidisc(points)
45  time2 = time.time()
46  d1t = time2-time1
47
48  d2 = onecenter2(points)
49  time3 = time.time()
50  d2t = time3-time2
51
52  d3 = onecenter1(points)
53  time4 = time.time()
54  d3t = time4-time3
55
56  print "Data preparation", time1-time0
57
58  print "Welzl              ", d1t, d1
59  print "Elzinga-Hearn      ", d2t, d2
60  print "Chrystal-Peirce    ", d3t, d3
```

The following is the output from one of the many runs. In each line the first number is the time taken to complete the search using a specific algorithm. While this is only from one particular run, it is consistent with the many runs that are done on my computer. Generally, we can see that without using data preparation, the Chrystal–Peirce algorithm takes more time than the other two. Data preparation using the Chakraborty–Chaudhuri algorithm only takes a fraction of the time used by the three algorithms, but it can significantly reduce the overall time used by each algorithm as shown in the last three lines. Also note that this example once again shows that the Chrystal–Peirce algorithm cannot find the optimal circle without using data preparation methods.

```
Welzl              0.856627941132  ((0.5, 0.5), 0.704372949832)
Elzinga-Hearn      0.605397939682  ((0.5, 0.5), 0.704372949832)
Chrystal-Peirce    0.903444051743  ((0.5, 0.5), 0.703718274146)
Data preparation 0.0939598083496
Welzl              0.264666080475  ((0.5, 0.5), 0.704372949833)
Elzinga-Hearn      0.301491975784  ((0.5, 0.5), 0.704372949832)
Chrystal-Peirce    0.237596035004  ((0.5, 0.5), 0.704372949832)
```

11.2 Location problems

The 1-center location problem is a special case of spatial optimization in the sense that a solution to this problem only contains one location. For many spatial optimization problems, we are more frequently in a situation where multiple locations must be found for purposes such as placing facilities (fire stations, stores, and libraries) at these locations.

In this section, we focus on a particular type of such optimization problems called the p-median problem, where the aim is to place p facilities in a network of n nodes such that the total distance from a location to its nearest facility location is minimized. Each of the n nodes is a demand node, and we aim to choose p nodes from them to place our facilities. To exactly solve such a problem, let us first mathematically formulate it as a mixed integer program. We refer the to locations in the network as nodes or vertices (we will use these two terms interchangeably) and use d_{ij} to denote the distance between the ith and jth nodes – we will use i to indicate the facility and j the demand nodes. There is a demand for the service from the facility at each node and we denote that as a_i for the ith node. To determine whether a facility is located at a node, we define a decision variable x_{ij} that takes the value 1 if demand at vertex i is assigned to the facility at vertex j and 0 otherwise, and x_{jj} that equals 1 if vertex j is a facility vertex that also serves itself and 0 otherwise. Using this notation, we write the following mixed integer program for the p-median problem:

$$\min \sum_{i=1}^{n} \sum_{j=1}^{n} a_i d_{ij} x_{ij}, \tag{11.1}$$

$$\text{subject to} \sum_{j=1}^{n} x_{ij} = 1 \quad \forall i, \tag{11.2}$$

$$\sum_{j=1}^{n} x_{jj} = p, \tag{11.3}$$

$$x_{ij} - x_{jj} \leq 0 \quad \forall \ i,j \text{ and } i \neq j, \tag{11.4}$$

$$x_{ij} = 0,1 \quad \forall \ i,j. \tag{11.5}$$

Here we have in (11.1) an objective function that minimizes the total distance weighted by demand. Each of the constraints in equation (11.2) states that, among all the selected nodes, demand at node i is assigned to one and only one selected node.

The constraint in equation (11.3) specifies that the total number of nodes selected for facilities is exactly p. The constraint in equation (11.4) guarantees that if demand at i is assigned to a facility at j (hence $x_{ij} = 1$), there must be a facility located at node j (hence $x_{jj} = 1$) so that $x_{ij} - x_{jj} \leq 0$. And if there is no facility at node j ($x_{jj} = 0$), then we should not assign the demand at node i to j ($x_{ij} = 0$), which again leads to $x_{ij} - x_{jj} \leq 0$. There are two more cases (see the exercises at the end of this chapter) to complete the description of equation (11.4). Finally, we ensure each decision variable (x_{ij} and x_{jj}) is a binary that takes a value of either 0 or 1.

The standard formulation of the p-median problem makes it possible to use fully fledged solvers to solve the problem. There are commercial solvers such as GUROBI and CPLEX that are proven to be effective. Additionally there are open source solvers that can be used, and in this chapter, as with all the chapters in this book, we rely on an open source tool called lp_solve.[3] While one may argue that the performance of open source tools has not been fully tested, they are adequate for our purposes of demonstrating the methods.

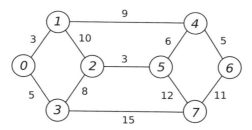

Figure 11.2 An example network with nodes labeled with italic numbers and edges marked with their lengths (not to scale)

We again use the example network used for the network analysis in Chapter 10 (Figure 11.2). To use our solver, we will need to convert the problem at hand into an input file that can be read by the solver. We use a popular file format called the LP format. The syntax of the LP file format is straightforward: we need to literally write every objective function and constraint in the formulation. An important task is to make sure all the decision variables are uniquely named. In our simple network case, we write decision variable x_{ij} as a string of the form x_i_j (e.g., x_5_16 for $i = 5$ and $j = 16$).

The Python program presented in Listing 11.6 can be used to write the LP file for our example p-median problem. Here we will once again utilize the all pair shortest path algorithm (Listing 10.7, imported in line 6) to compute the distance matrix d_{ij} in the p-median problem formulation. Line 19 reads the raw input file that only contains the adjacent edges into a matrix form and line 20 computes the distance matrix. We assume the demand is 1 for all nodes in the network. The code block from line 23 creates the string for the decision variables. Strings are used to store the objective function and each of the constraints. The entire LP file is then printed at the end of the program.

---

[3] http://lpsolve.sourceforge.net/5.5/

Listing 11.6: Writing an LP file for the *p*-median problem (pmed_lpformat.py).

```
1   import sys
2   from string import atoi
3   import sys
4   sys.path.append('../networks')
5   from network2listmatrix import network2distancematrix
6   from allpairdist import *
7
8   if len(sys.argv) <=1:
9       print sys.argv[0], "filename [ [Ture|False] p]"
10      sys.exit()
11
12  zerobased = True
13  p = 2
14  if len(sys.argv)>=3 and sys.argv[2]=="False":
15      zerobased = False
16  if len(sys.argv)>=4:
17      p = atoi(sys.argv[3])
18
19  a = network2distancematrix(sys.argv[1], zerobased)
20  allpairs(a)
21  n = len(a)
22
23  X = [['']*n for i in xrange(n)]
24  for i in xrange(n):
25      for j in xrange(n):
26          X[i][j] = 'x_%d_%d'%(i, j)
27
28  # objective function
29  obj='MIN: '
30  for i in range(n):
31      for j in range(n):
32          obj += "%d"%(a[i][j])+"*"+X[i][j] # %f if real numbers
33          if not (i == n-1 and j == n-1):
34              obj += '+'
35          else:
36              obj=obj+';'
37
38  # constraints
39  con1 = ['' for i in range(n)]
40  for i in range(n):
41      con1[i] = X[i][0]
42      for j in range(1,n):
43          con1[i] += '+'+X[i][j]
44      con1[i] += '=1;'
45
46  con2 = X[0][0]
47  for j in range(1, n):
48      con2 += '+'+X[j][j]
49  con2 += '=%d;'%(p)
```

```
50
51  con3 = []
52  for i in range(n):
53      for j in range(n):
54          if i <> j:
55              con3.append(X[i][j] + '-' + X[j][j] + '<=0;')
56
57  con4 = 'BIN '
58  for i in range(n):
59      for j in range(n):
60          con4 += X[i][j]
61          if not (i == n-1 and j == n-1):
62              con4 += ','
63          else:
64              con4 += ';'
65
66  # print it out
67  print obj
68  for i in range(n):
69      print con1[i]
70  print con2
71  for i in range(len(con3)):
72      print con3[i]
73  print con4
```

Listing 11.7: A portion of the LP file for the simple network.

```
1  MIN: 0*x_0_0+3*x_0_1+13*x_0_2+5*x_0_3+...+11*x_7_6+0*x_7_7;
2  x_0_0+x_0_1+x_0_2+x_0_3+x_0_4+x_0_5+x_0_6+x_0_7=1;
3  ...
4  x_0_0+x_1_1+x_2_2+x_3_3+x_4_4+x_5_5+x_6_6+x_7_7=2;
5  x_0_1-x_1_1<=0;
6  ...
7  x_7_6-x_6_6<=0;
8  BIN x_0_0,x_0_1,x_0_2,x_0_3,...,x_7_6,x_7_7;
```

Listing 11.7 shows a very small portion of the result of running the Python program on the simple network data. The two lines of commands below illustrate how the Python program is used to create the LP file and the lp_solve software is used to solve the problem in the LP file. The result suggests that when p is 2 (the default value in our Python program), it is optimal to locate the facilities at nodes 0 and 5 since x_0_0 and x_5_5 are 1. The result also suggests which of the two facilities should be assigned to each of the nodes. For example, node 2 should be served by the facility at node 5, as the value of x_2_5 is 1. Examining the network in Figure 11.2, this is reasonable because node 2 is closer to node 5 than to node 0.

```
$ python pmed_lpformat.py network-links > network-simple.lp
$ lp_solve -ia -S0 network-simple.lp
```

```
Value of objective function: 40.00000000

Actual values of the variables:
x_0_0                    1
x_1_0                    1
x_2_5                    1
x_3_0                    1
x_4_5                    1
x_5_5                    1
x_6_5                    1
x_7_5                    1
```

We now test the Python program in Listing 11.6 on a problem from a set of benchmark problems, the 40 p-median problems in the OR-Library.[4] Each of these problems is stored in a text file that is very similar to what we used in Listing 10.1: the first line contains three items instead of one as in our case. But Python will still work because it only reads the first number, which is the number of total nodes in the network. The first of the 40 problems has 100 nodes and the p value is 5. Suppose the file for this problem is called pmed1.orlib. We run our program as follows and it returns the correct objective function value. Note that we need to specify that the file is not zero based and the p value is 5 so we can compare the result obtained with the known objective function value.

```
$ python pmed_lpformat.py pmed1.orlib False 5 > pmed1.lp
$ lp_solve -S1 pmed1.lp
Value of objective function: 5819.00000000
```

11.3 Notes

According to Sylvester (1860), Peirce developed the first algorithm solving the minimal enclosing circle problem. Later, Chrystal (1885) formalized Peirce's algorithm. Recent theoretical analysis has shown that the Chrystal–Peirce algorithm has a time complexity of $O(n^2)$ (Chakraborty and Chaudhuri, 1981; Plastria, 2002). The start strategy for the Chrystal–Peirce algorithm was developed by Chakraborty and Chaudhuri (1981). The Elzinga–Hearn algorithm was developed by Elzinga and Hearn (1972), and an analysis by Drezner and Shelah (1987) showed that the overall iterations required to complete this algorithm are $O(n^2)$ and the time complexity is $\Omega(n^2)$. For the algorithm by Welzl (1991), the expected time complexity of this algorithm is $O(n)$ (see also de Berg et al., 1998, p. 89).

---

[4]http://www.brunel.ac.uk/~mastjjb/jeb/info.html. The direct link to the 40 uncapacitated p-median problems used in this book is http://people.brunel.ac.uk/~mastjjb/jeb/orlib/pmedinfo.html.

11.4 Exercises

1. The `test` function in the Chrystal–Peirce algorithm code outputs points where each coordinate is a floating point number with a limited precision of one decimal point. Because of that, the output listed in the text is not useful in drawing the points and helping us understand what is the condition that causes the initial circle not to be optimal. Modify the code so that only integers are used to create the test points. Then try to run the program many times until the first enclosing circle reported is not optimal. Draw those points and discuss why the initial circle is not optimal.

2. The example given in Figure 11.1 always uses an obtuse triangle during the search process. Since a random point outside the circle is chosen at each iteration, try to change the illustration so that at least one acute triangle is used in the process of finding the optimal enclosing circle.

3. We discussed two cases for constraints (11.4) in the p-median problem formulation. What are the other two cases that this set of constraints is designed to guarantee? How can these constraints help make sure that each demand is assigned to its nearest facility?

4. The 1-center problem discussed in this chapter has a multi-facility version called the p-center problem. The continuous version of the center problem is meant to locate p facilities so that each one sits at the center of the points that can be reached. We require the maximum distance from a center to any points assigned to it to be minimized. This is also called a minimax problem because we are minimizing the maximum distance. The discrete version of the p-center problem requires a set of the candidate locations where the facilities can be placed. To be able to compute the distances, we use d_{ij} to denote the distance between vertices i and j, where i belongs to a set I of the demand locations and j belongs to the set J of potential locations for facilities. The decision variables include y_j, which equals 1 if vertex j is selected or 0 otherwise; x_{ij}, which equals 1 if demand at i is served by the facility at j; and z, the minimum distance between a vertex and its nearest facility location. We then formulate the problem as follows:

$$\min z,$$

$$\text{subject to } \sum_j x_{ij} = 1 \quad \forall i,$$

$$\sum_j y_j = p,$$

$$x_{ij} \leq y_j \quad \forall i,j,$$

$$z \geq \sum_j d_{ij} x_{ij} \quad \forall i,$$

$$y_j = 0,1 \quad \forall j,$$

$$x_{ij} = 0,1 \quad \forall i,j.$$

Explain the meaning of each line in this formulation and write a Python program to output an LP file for the *p*-center problem using our simple network data.

5. While we have only focused on the median and center problems, there is a big family of location problems called covering problems that have had many applications. Suppose we have a set of demand points and each of them is denoted by *i*. We need to place facilities at a set of candidate locations, and each of the candidate locations is denoted by *j*. The cost of placing a facility at location *j* is f_j. Our aim is to make sure that all the demand points are covered by at least one of the facilities. Being covered can have various meanings. For example, if we build a watch tower at location *j* and the demand point at *i* can be watched from that tower, we say *i* is covered by *j*. We use a variable a_{ij} to denote this kind of coverage relationship, and its value is 1 if the facility at *j* can cover demand at *i*, otherwise we set the value to 0. We use a set N_i to denote the indices of potential facilities that can cover demand at *i*. Formally, we have $N_i = \{j | a_{ij} = 1\}$. The decision variable is x_j, which is 1 if a facility is located at candidate location *j*, and 0 otherwise. With this notation, the set covering problem can be formulated as follows:

$$\min \sum_{j=1} f_j x_j,$$

$$\text{subject to } \sum_{j \in N_i} x_j \geq 1 \quad \forall i,$$

$$x_j = 0, 1 \quad \forall j.$$

Write a Python program that can be used to write the set covering problem in the LP format for lp_solve to solve.

12

Heuristic Search Algorithms

"Would you tell me, please, which way I ought to go from here?"
"That depends a good deal on where you want to get to," said the Cat.
"I don't much care where—" said Alice.
"Then it doesn't matter which way you go," said the Cat.
'—so long as I get *somewhere*," Alice added as an explanation.
"Oh, you're sure to do that," said the Cat, "if you only walk long enough."

Lewis Carroll, *Alice's Adventures in Wonderland*

Now we shift the focus of our discussion toward heuristic methods, the methods that are generally efficient but cannot guarantee optimal solutions to be found. There are actually two kinds of heuristics we will talk about: those designed for specific problems and those for a wide range of general problems. The latter is called metaheuristics because they are not just for a single type of problem but are suitable for many kinds of problems. These algorithms are often called search algorithms because their purpose is to search through the solution space with the aim of finding the best one.

12.1 Greedy algorithms

We can construct a feasible solution to a problem from scratch. For example, to find a solution to the p-median problem on a network, we can start by first adding a facility node into a partial solution and keep adding new nodes until a complete solution is formed. There must be criteria that guide the way of choosing which node is added each time. Each time a new node is added into the partial solution, the sum of distances from each node to its nearest facility node will decrease. Because of that, we can come up with a rule of thumb such that each time we only add the node that yields the largest decrease in the total distance. We can also start from a "super" solution where all the nodes are selected and the objective function value, if we have to compute it, will be zero. Then we start to drop nodes from the solution until a feasible solution (i.e., with p nodes selected) is reached.

These are called greedy algorithms because we only take what is considered to be the best option at each step. In general, we design a greedy algorithm for a problem by incrementally managing a *partial* solution until a complete solution is constructed. During this process, we should know the components of a feasible solution and how to choose each component toward the whole solution. The aim of a greedy algorithm is to obtain *one* solution. Therefore we need to make sure only feasible solutions will be obtained using the algorithm. Along the way to obtaining the solution, we need to evaluate each of the partial solutions using the objective function.

The Python program in Listing 12.1 implements a greedy algorithm for the *p*-median problem using the addition principle. The function `evaluate` (line 9) returns the sum of distances from each node to its nearest facility given the solution in `median`. Here, we use the all pair shortest path algorithm again to obtain the distance matrix from a *p*-median file that contains information for each edge. The basic design idea of this algorithm is to start with an empty list of nodes (line 34) and in each iteration use a `for` loop to test which node should be added into the solution so that the total distance of the solution is minimal (line 39). Once such a node is found, we remove it from the candidates (line 44) so that it will not be considered again in future iterations. We also add the node into the current solution (line 45).

Listing 12.1: A greedy algorithm for the *p*-median problem (pmed_greedy.py).

```python
import sys
from string import atoi
sys.path.append('../networks')
from network2listmatrix import network2distancematrix
from allpairdist import allpairs

INF = float('inf')

def evaluate(dist, median, n):
    sumdist = 0.0
    p = len(median)
    for i in range(n):
        dist0 = INF
        for j in range(p):
            if dist[i][median[j]] < dist0:
                dist0 = dist[i][median[j]]
        sumdist += dist0
    return sumdist

if len(sys.argv) <=1:
    print sys.argv[0], "filename [ [True|False] p]"
    sys.exit()

zerobased = True
p = 2
if len(sys.argv)>=3 and sys.argv[2]=="False":
    zerobased = False
if len(sys.argv)>=4:
    p = atoi(sys.argv[3])

a = network2distancematrix(sys.argv[1], zerobased)
allpairs(a)
```

```
33  n = len(a)
34  median = []
35  candidates = [i for i in range(n)]
36  for j in range(p):
37      dmin = INF
38      imin = -1
39      for i in candidates:
40          d = evaluate(a, median+[i], n)
41          if d < dmin:
42              dmin = d
43              imin = i
44      candidates.remove(imin)
45      median.append(imin)
46
47  print median, dmin
```

We test the greedy algorithm on the two p-median problems we discussed in the previous section. The result shows that we can find a solution quickly, though it is not optimal. At this point, it should be clear to us that greedy algorithms are quick and simple. They probably do not mean to be perfect in finding the exact optimal solutions for many cases, but they can find feasible solutions quickly.

```
$ python pmed_greedy.py ../data/network-links
[2, 0] 49.0

$ python pmed_greedy.py ../data/pmed1.orlib False 5
[6, 12, 3, 90, 98] 5891.0
```

12.2 Vertex exchange algorithm

The solutions to the p-median problems found by the greedy algorithm appear to be non-optimal, for both of the example problems used in the previous section. We now introduce a heuristic algorithm that is designed specifically for the p-median problem. This is the vertex exchange algorithm, often referred to as the Teitz–Bart algorithm after the authors who first developed it. The idea here is simple: we start from a random solution and keep exchanging one of the selected vertices with an unselected vertex until we cannot further improve the solution. We write the logic here in pseudo-code in Listing 12.2.

Listing 12.2: The vertex exchange algorithm for the p-median problem.
```
1  Let s be a random solution
2  Repeat
3      For each unselected vertex i
4          Get the gain by replacing a selected vertex with i
5      Let s1 be the solution with the best gain
6      If the gain of s1 > 0, then s = s1
7  Until gain of s1 < 0
```

The Python program in Listing 12.3 implements the vertex exchange algorithm. A solution to the problem here is represented using a string of p integers, each indicating the index of the selected vertex. A key process in the algorithm is to test an alternative solution and evaluate whether there is a net gain in the objective function value in the new solution. We always create an alternative solution by inserting a new, unselected vertex in the current solution and test which selected vertex can be replaced by the new one (line 4 in the pseudo-code). To find the best replacement, one can simply calculate the nearest selected vertex to each vertex and then find the total distance by summation. While this is doable, it is not efficient.

Here, we introduce an algorithm that does not recompute everything. This is implemented in the function called `findout` (line 11). In this function, instead of recomputing, we utilize two lists of assignments given a solution of p selected facility vertices. The first list (`d1`) stores the index of the nearest facility to each vertex, while the second list stores the second nearest facility to each vertex. In inserting a new vertex into the current solution, we must take away one of the existing vertices, and we want to do this so that we can gain as much as possible in terms of the objective function value. If a net gain is impossible, then we want the loss in objective function value to be minimal. We proceed by going through all the vertices and testing for a potential replacement. The vertex we test for insertion is named `fi`. For each vertex `i` (line 19), if the distance between `i` and `fi` is smaller than the distance between `i` and its current nearest, then a replacement would be a good deal and we accumulate the gain (line 21). Otherwise, we are looking at a potential loss if we have to remove the current nearest facility `i`, and the loss is dependent on which node is now closer to `i`, the second nearest facility or the new vertex to be inserted (line 24). We accumulate the loss from removing each current facility vertex (line 24). We will choose the current vertex with minimal loss (line 27) and calculate the total gain (line 31).

This algorithm has proved to be extremely efficient because we successfully exclude cases where the reallocation has to be made for all the vertices when inserting a new one into the solution. This is a typical strategy of tradeoff between space and time: we use more storage to gain computation time. The downside of this is that we will need to maintain the lists `d1` and `d2`. This is done in the function `update_assignment` (line 34), where the first and second nearest facility vertices are checked for each vertex in the data.

The `next` function (line 59) finds the best replacement after checking all the vertices to be inserted into the current solution (line 73). If a better solution is found (line 82), we make the replacement and actually do the update and assignment. The first returned value is a Boolean that tells when a better solution is found or not found.

The `teitz_bart` function (line 88) drives the whole iterative process of the vertex exchange algorithm as listed in the pseudo-code. We first create an initial solution by randomly drawing p vertices from the N vertices in the network (line 96). The initial solution is evaluated using the function `update_assignment` (line 99). A `while` loop is then used to iteratively test whether exchanging vertices can improve the solution.

Listing 12.3: Vertex exchange algorithm for the *p*-median problem (teitz_bart.py).

```
1  from bisect import bisect_left
2  from string import atoi, split
3  import random
4  import sys
```

```python
sys.path.append('../networks')
from network2listmatrix import network2distancematrix
from allpairdist import allpairs

INF = float('inf')

def findout(median, fi, dist, d1, d2, N):
    """
    Determines, given a candidate for insertion (fi),
    the best candidate in the solution to replace or remove
    (fr).
    This does not change values in median, d1, and d2.
    """
    w = 0.0
    v = [0.0 for i in range(N)]
    for i in range(N):
        if dist[i][fi] < dist[i][d1[i]]:                  # gain
            w += dist[i][d1[i]] - dist[i][fi]
        else:                                              # loss
            v[d1[i]] += min(dist[i][fi],
                            dist[i][d2[i]]) - dist[i][d1[i]]
    fmin = INF
    fr = 0
    for i in median:
        if v[i] < fmin:
            fmin = v[i]
            fr = i
    fmin = w-fmin
    return fmin, fr # gain and vertex to be replaced

def update_assignment(dist, median, d1, d2, p, N):
    """
    Updates d1 and d2 given median so that d1 holds the
    nearest facility for each node and d2 holds the second
    """
    dist1, dist2 = 0.0, 0.0
    node1, node2 = -1, -1
    for i in range(N):
        dist1, dist2 = INF, INF
        for j in range(p):
            if dist[i][median[j]] < dist1:
                dist2 = dist1
                node2 = node1
                dist1 = dist[i][median[j]]
                node1 = median[j]
            elif dist[i][median[j]] < dist2:
                dist2 = dist[i][median[j]]
                node2 = median[j]
        d1[i] = node1
        d2[i] = node2
    dist1 = 0
    for i in range(N):
        dist1 += dist[i][d1[i]]
```

```
57          return dist1
58
59   def next(dist, median, d1, d2, p, N):
60          """
61          INPUT
62            dist: distance matrix
63            median: list of integers for selected vertices
64            d1: list of nearest facility for each vertex
65            d2: list of second nearest facility
66            p: number of facilities to locate
67            N: number of vertices on the nextwork
68          OUTPUT
69            r: total distance
70            median: list of integers for selected vertices
71          """
72          bestgain = -INF
73          for i in range(N):
74              gain, fr1 = findout(median, i, dist, d1, d2, N)
75              if i in median:
76                  continue
77              if gain>bestgain:
78                  bestgain = gain
79                  fr = fr1
80                  fi = i
81          r = 0
82          if bestgain > 0:
83              i = median.index(fr)
84              median[i] = fi
85              r = update_assignment(dist, median, d1, d2, p, N)
86          return bestgain>0, r, fr, fi
87
88   def teitz_bart(dist, p, verbose=False):
89          """
90          INPUT
91            dist: distance matrix
92            p: number of facilities to be selected
93            verbose: whether intermediate results are printed
94          """
95          N = len(dist)
96          median = random.sample(range(N), p)
97          d1 = [-1 for i in range(N)]
98          d2 = [-1 for i in range(N)]
99          r = update_assignment(dist, median, d1, d2, p , N)
100         if verbose: print r
101         while True:
102             result = next(dist, median, d1, d2, p, N)
103             if result[0]:
104                 r = result[1]
105                 if verbose: print r
106             else:
```

```
107                break
108        return r, median
109
110 if __name__ == "__main__":
111     print 'Problem: simple network'
112     a = network2distancematrix('../data/network-links', True)
113     allpairs(a)
114     teitz_bart(a, 2, True)
115     print 'Problem: pmed1 in OR-lib'
116     a = network2distancematrix('../data/orlib/pmed1.orlib',
117                                 False)
118     allpairs(a)
119     teitz_bart(a, 5, True)
```

Similar to what we did with the greedy algorithm, we test the vertex exchange algorithm using the simple network (line 114) and the first *p*-median problem in the OR-Library (line 119). The result given below shows that we are able to find the optimal solutions to both problems. After many reruns, the program always seems to find the optimal solutions to these two test examples.

```
Problem: simple network
53
43
40
Problem: pmed1 in OR-lib
11030
7971
6598
6024
5847
5821
5819
```

How about other problems? Let us carry out a test using more benchmark problems. This time we will use all 40 problems in the OR-Library and run each problem only once (for the sake of illustration only). Some of the problems will be tested many times, and it will become more convenient and efficient if we only convert the network file stored as a list into a distance matrix once. We now save all the distance matrices beforehand and load them into the problem each time we run them. This can be easily done using the `list2distancematrix` function discussed in the context of network analysis (Chapter 10), saving each of the distance matrices in a text file. The program in Listing 12.4 includes a function called `load_distance_matrix` that loads a distance matrix with a specified file name. The specifications of the 40 problems are listed in the variables starting at line 21. The function `testall` runs each of the problems once and function `testpmed` tests a specified problem 20 times by default.

Listing 12.4: Testing problems in the OR-Library using the vertex exchange algorithm (test_orlib_pmed.py).

```python
import time
from teitz_bart import *

def load_distance_matrix(fname):
    # get distance matrix
    f = open(fname)
    l = f.readline()
    N = atoi(split(l.strip())[0]) # get the number of nodes
    dist=[[0]*N for x in xrange(N)] # 2D list filled with INF
    l = f.readline()
    row = 0
    for l in f:
        dline = split(l.strip())
        if len(dline)==N:
            for i in range(N):
                dist[row][i] = atoi(dline[i])
            row += 1
    f.close()
    return dist

N = [100, 100, 100, 100, 100, 200, 200, 200, 200, 200,
     300, 300, 300, 300, 300, 400, 400, 400, 400, 400,
     500, 500, 500, 500, 500, 600, 600, 600, 600, 600,
     700, 700, 700, 700, 800, 800, 800, 900, 900, 900]
pp = [5, 10, 10, 20, 33, 5, 10, 20, 40, 67,
      5, 10, 30, 60, 100, 6, 10, 40, 80, 133,
      5, 10, 60, 100, 167, 5, 10, 60, 120, 200,
      5, 10, 70, 140, 5, 10, 80, 5, 10, 90]
known = [5819, 4093, 4250, 3034, 1355,
         7824, 5631, 4445, 2734, 1255,
         7696, 6634, 4374, 2968, 1729,
         8162, 6999, 4809, 2845, 1789,
         9138, 8579, 4619, 2961, 1828,
         9917, 8307, 4498, 3033, 1989,
         10086, 9297, 4700, 3013, 10400,
         9934, 5057, 11060, 9423, 5128]

def testall():
    for i in range(len(N)):
        fn = format("../data/orlib/pmed%d.distmatrix"%(i+1))
        dist = load_distance_matrix(fn)
        t1 = time.time()
        r = teitz_bart(dist, pp[i])[0]
        t2 = time.time()
        print format("pmed%d:"%(i+1)),\
            N[i], pp[i], r, known[i], t2-t1,
        if r==known[i]: print "(*)"
```

```
            else: print
def testpmed(i, numtest=20):
    fn = format("../data/orlib/pmed%d.distmatrix"%(i+1))
    dist = load_distance_matrix(fn)
    t1 = time.time()
    sumdiff = 0
    nummiss = 0
    for j in range(numtest):
        r= teitz_bart(dist, pp[i], verbose=True)[0]
        print r,
        if r==known[i]: print "(*)"
        else:
            print
            sumdiff += r - known[i]
            nummiss += 1
    t2 = time.time()
    print "Time:", (t2-t1)/numtest
    print "Found:", numtest-nummiss
    if nummiss:
        print "Avg diff:", float(sumdiff)/nummiss

if __name__ == "__main__":
    testall()
    testpmed(16)
```

The results of running the `testall` function are given below. It is clear that the vertex exchange algorithm can indeed find optimal solutions to many problems: for this experiment 14 out of 40, to be precise. It is interesting to note that among different p values for a certain n value, the algorithm tends to find optimal solutions when p is relatively small. These are "easier" problems in that setting (when n is given) because they have fewer possible combinations. In each line of the output below, we print the name of the problem (e.g., pmed1), the total number of vertices (n), number of facilities to be located (p), the best objective function value found, the known optimal solution, and time taken (in seconds). An asterisk is used to indicate that the optimal solution is found.

```
pmed1:  100  5 5819 5819 0.0466809272766 (*)
pmed2:  100 10 4102 4093 0.0773859024048
pmed3:  100 10 4250 4250 0.0820319652557 (*)
pmed4:  100 20 3049 3034 0.122255086899
pmed5:  100 33 1355 1355 0.150933027267 (*)
pmed6:  200  5 7824 7824 0.213330984116 (*)
pmed7:  200 10 5639 5631 0.414077043533
pmed8:  200 20 4454 4445 0.76794219017
pmed9:  200 40 2764 2734 1.02096009254
pmed10: 200 67 1265 1255 1.57539701462
pmed11: 300  5 7696 7696 0.4705991745 (*)
pmed12: 300 10 6634 6634 1.02756404877 (*)
```

```
pmed13:  300  30   4374  4374  1.90632295609  (*)
pmed14:  300  60   2971  2968  4.46003508568
pmed15:  300  100  1737  1729  5.15211892128
pmed16:  400  6    7788  8162  1.28738808632
pmed17:  400  10   7014  6999  1.3513071537
pmed18:  400  40   4813  4809  5.61389303207
pmed19:  400  80   2860  2845  8.77476096153
pmed20:  400  133  1805  1789  9.67293000221
pmed21:  500  5    9138  9138  1.22582006454  (*)
pmed22:  500  10   8579  8579  2.59184718132  (*)
pmed23:  500  60   4212  4619  13.2033700943
pmed24:  500  100  2962  2961  17.4130480289
pmed25:  500  167  1844  1828  23.4592969418
pmed26:  600  5    9917  9917  1.53462600708  (*)
pmed27:  600  10   8307  8307  3.74222993851  (*)
pmed28:  600  60   4508  4498  15.5038609505
pmed29:  600  120  3041  3033  24.4620049
pmed30:  600  200  2012  1989  32.912113905
pmed31:  700  5    10087 10086 2.17640995979
pmed32:  700  10   9301  9297  4.40583491325
pmed33:  700  70   4715  4700  24.5156989098
pmed34:  700  140  3025  3013  41.5150740147
pmed35:  800  5    10400 10400 3.16665506363  (*)
pmed36:  800  10   9951  9934  7.37093496323
pmed37:  800  80   5067  5057  38.7697281837
pmed38:  900  5    11060 11060 5.15535211563  (*)
pmed39:  900  10   9423  9423  7.04349589348  (*)
pmed40:  900  90   5134  5128  56.506018877
```

Now the question is, can this algorithm actually find optimal solutions to any problem, regardless of how many tries are needed? Clearly there are many problems to which the algorithm cannot find the optimal solution with a single shot. For example, in the above run, problem pmed17 has an optimal objective function value of 6,999, but the algorithm only returns 7,014. Is there a hope that we can find the optimal solution here? Let us try the algorithm more times, say 20, for this particular case. This is done in the function called `testpmed` by specifying the number of the problem (note the number here starts from zero). As shown below, it turns out that the algorithm can find the optimal solution in 10 out of 20 attempts.

```
6999 (*)
6999 (*)
7037
6999 (*)
6999 (*)
7003
7010
6999 (*)
6999 (*)
6999 (*)
```

```
7014
7037
6999 (*)
7014
7014
6999 (*)
6999 (*)
7003
7014
7003
Time: 1.27441074848
Found: 10
Avg diff: 15.9
```

12.3 Simulated annealing

The vertex exchange algorithm for the p-median problem is "greedy"[1] because it considers the best replacement in each iteration and only those that improve the solution are accepted. This makes it pretty easy to trap the search process into some kind of local optimal solution where the next possible solution is bound to be worse than the current one. What if we give worse solutions a chance? Humans are greedy but are also strategic. In chess, for example, we commonly sacrifice pieces of lower value to achieve the ultimate aim of checkmating the opponent. This type of thinking leads to a variety of new heuristic approaches.

In this section, we introduce a new heuristic method called simulated annealing. Similar to other heuristic methods, a simulated annealing process needs a strategy to create a set of alternative solutions to consider at each iteration. These new solutions are generally referred to as the neighbors of the current solution, because they are typically created using a set of operations to change the current solution. Once the neighbors have been created, we apply a selection strategy so that one is picked. We can pick the best among the neighbors, but sometimes a random solution from the neighbors can also be used. Then we need to decide whether the new solution is to be accepted as the next solution. This depends on how worse solutions are handled and how the algorithm is designed to escape local traps in order to reach optimal solutions. Specifically, when a worse solution is encountered in the search process, there is a chance that it will be accepted, especially at the beginning of search. Given a new solution, we use Δ to denote the difference between its objective function from the current solution. Suppose that we want to minimize the objective function and a negative Δ value means an improvement over the current solution and we will accept it for the next round. A worse solution still has a probability of being accepted, and we calculate this probability using the Metropolis algorithm:

$$\Pr(\Delta, T) = \begin{cases} 1 & \text{if } \Delta < 0 \\ e^{-\Delta/T} & \text{otherwise,} \end{cases}$$

---
[1] It is not accurate to call it a greedy algorithm though, because this name is reserved for another category of algorithms.

where T is a parameter that is often referred to as temperature. During the search process, the value of T decreases with the iterations. Clearly a high T value will give worse solutions a high probability. We use an analog of annealing to describe this decreasing temperature as happens in a physical annealing process. The strategy of decreasing temperature is called the cooling schedule. A simple way to decrease temperature is to multiply T by a rate r ($0 < r < 1$) at the end of each iteration. Once the acceptance probability is calculated, we can draw a random sample from a uniform distribution between 0 and 1. If the sample is smaller than the probability then we accept the new solution. Otherwise we reject the new solution and stop the process.

Listing 12.5 shows pseudo-code for a general simulated annealing procedure. We assume the objective function value of a solution s can be calculated as $f(s)$, and $R(0,1)$ is a function that returns a random number drawn from a uniform distribution. The algorithm continues to accept new solutions until the condition in line 9 is not met.

Listing 12.5: The general simulated annealing algorithm.

```
1   Let s be an initial solution to the problem
2   Let f(s) be the objective function value of s
3   Set T to an initial value
4   Repeat
5       Find a set of neighbors of s
6       Let s1 be the best solution in the neighbors
7       Δ = f(s1) - f(s)
8   P = Pr(Δ, T)
9   If P > R(0, 1)
10          s = s1
11          T = r * T
12  Else
13          Stop and report the best solution found
```

We now implement a simulated annealing algorithm for the p-median problem (Listing 12.6). The core function `simulated_annealing` (line 117) drives the flow of the overall procedure as outlined in the pseudo-code. Similar to the vertex exchange algorithm, we initialize the solution as a set of random indices (line 125). The same `update_assignment` used before is also used here (line 126). We continue to use the idea of nearest and second nearest facilities for each vertex here to make it efficient to test vertex exchange. We set the initial temperature to 100 (line 132). In the `while` loop (line 133), we call the function `next` (line 134) to find the best solution for the next iteration.

The general procedure in the function `next` is to find the best solution in the neighbors of the current solution and test if the new solution should be accepted. We have implemented two ways of creating the neighbors of a given solution and therefore we have a choice of which one to use. The choice is specified as a parameter called `neighbor-method` of the `next` function. The first choice is to create a neighbor by replacing one of the selected vertices with another vertex that is within a certain distance as specified by the `dthreshold` parameter. We create a total of p neighbors using a function called `bestGeoNeighbor` (line 64), which returns the best neighbor for acceptance testing later in the `next` function (line 89). Note that function `bestGeoNeighbor` does not return an actual solution; instead it returns the replacement of a current selected vertex (fr) with a new vertex (fi). Because this way of creating neighbors is based on replacing a vertex in an existing solution, there is no need to recompute all the assignments and we utilize the nearest and second nearest facilities of each vertex (line 74), the same mechanism used in the findout function introduced before. The second approach is to generate *p* random new solutions and the best one is returned, as implemented in function bestRandomNeighbor (line 50). Since each neighbor solution here is a new one, there is a need to calculate the objective function value anew (line 55) and there is no need to maintain the nearest and second nearest lists.

There are a couple more tricks in the simulated annealing method implemented here. First, we use a variable called accepted_same (line 131) to record the times when the same best solution was accepted after a worse solution has been accepted (line 141). The second is to only take roughly two thirds of the replacements for each selected vertex (line 72). Both strategies are designed to prevent unnecessary oscillation between the best solution found and the second best solution found. We know it is in the nature of a simulated annealing that it may accept solutions worse than the best solution found. Suppose the algorithm has found the optimal solution, so that any alternative solution must be worse than the optimal. If we try a greedy approach so that only the best one of the neighbors can be chosen, the same worse solution will always be chosen, and then the next iteration will come back to the same best solution. Of course this can also happen on a local optimal solution. For this reason, we arbitrarily set the number of times to come back to the same best solution at three. This also helps to terminate the algorithm sooner because otherwise it will continue to run for more iterations until the acceptance probability is low enough to stop this process. By only allowing two thirds of the replacements to be chosen, we further reduce the chance of coming back to the same solution.

Listing 12.6: A simulated annealing algorithm (simulated_annealing.py).

```
from bisect import bisect_left
from string import atoi, split
import math
import random
from copy import deepcopy
from teitz_bart import update_assignment
import sys
sys.path.append('../networks')
from network2listmatrix import network2distancematrix
from allpairdist import allpairs

INF = float('inf')

def evaluate(dist, median, p, N):
    sumdist = 0.0
    for i in range(N):
        dist0 = INF
        for j in range(p):
            if dist[i][median[j]] < dist0:
                dist0 = dist[i][median[j]]
        sumdist += dist0
```

```
22          return sumdist
23
24      # test replacing fr with fi in median without reallocating
25      # all the nodes
26      def test_replacement(fi, fr, dist, d1, d2, p, N):
27          total = 0.0
28          for i in range(N):
29              if dist[i][fi]<dist[i][d1[i]]:
30                  total += dist[i][fi]
31              else:
32                  if fr==d1[i]:
33                      if dist[i][fi]<dist[i][d2[i]]:
34                          total += dist[i][fi]
35                      else:
36                          total += dist[i][d2[i]]
37                  else:
38                      total += dist[i][d1[i]]
39          return total
40
41      def get_new_median(N, p, median):
42          samples = []
43          for i in range(p):
44              s = -1
45              while s in median or s in samples:
46                  s = random.sample(range(N), 1)[0]
47              samples.append(s)
48          return samples
49
50      def bestRandomNeighbor(median, dist, N, p):
51          candidates = []
52          r_mins = []
53          for j in range(p):
54              candidates.append(get_new_median(N, p, median))
55              r_mins.append(evaluate(dist, candidates[j], p , N))
56          r_min = INF
57          i_min = -1
58          for j in range(p):
59              if r_mins[j] < r_min:
60                  r_min = r_mins[j]
61                  i_min = j
62          return r_min, candidates[i_min]
63
64      def bestGeoNeighbor(median, dist, d1, d2, N, p, dthreshold):
65          r_min = INF
66          for j in range(p):
67              r_min_temp = INF
68              fi_temp = -1
69              for i in range(N): # try replacing j with an i
70                  if dist[median[j]][i]>dthreshold or i in median:
71                      continue
72                  if random.random() > 0.67:
```

```
                    continue
                r1 = test_replacement(i, median[j], dist,
                                      d1, d2, p, N)
                if r1 < r_min_temp:
                    r_min_temp = r1
                    fi_temp = i
            if r_min_temp < r_min:
                r_min = r_min_temp
                fi = fi_temp
                fr = j
    return r_min, fi, fr

def next(r, T, median, dist, d1, d2, p, N,
         dthreshold, neighbormethod):
    r1 = r
    if neighbormethod == 0:
        r_min, fi, fr = bestGeoNeighbor(median, dist, d1,
                                        d2, N, p, dthreshold)
        test = acceptable(10*(r_min-r)/r, T)
        if test[0] > 0:
            median[fr] = fi
            r1 = update_assignment(dist, median,d1,d2,p,N)
        return test[0], r1, median
    else:
        r_min, candidate = bestRandomNeighbor(median,
                                              dist, N, p)
    test = acceptable(10*(r_min-r)/r, T)
    if test[0] > 0:
        median = candidate
        r1 = update_assignment(dist,
                               median, d1, d2, p, N)
    return test[0], r1, median

def acceptable(delta, T):
    if delta<0:                  # better solution
        return 1, 1
    if delta==0:                 # same solution, no change
        return 0, 0
    prob = math.exp(-delta/T)
    if random.random() < prob:   # worse solution, accept
        return 2, prob
    return 0, prob               # worse solution, reject

def simulated_annealing(dist, p,
                        neighbormethod=0, verbose=False):
    N = len(dist)
    dmax = max([max(dist[i]) for i in range(N)])
    dthreshold = dmax/2
    d1 = [-1 for i in range(N)]
    d2 = [-1 for i in range(N)]
```

```python
            ## Initialization
            median = random.sample(range(N), p)
            r = update_assignment(dist, median, d1, d2, p , N)
            first = [deepcopy(r), deepcopy(median)]
            best = [r, median]
            if verbose: print first[0]

            accepted_same = 0
            T = 100.0
            while True:
                result = next(r, T, median, dist, d1, d2, p, N,
                              dthreshold, neighbormethod)
                if result[0]>0:
                    r = result[1]
                    if r < best[0]:
                        best = [r, deepcopy(result[2])]
                        accepted_same = 0
                    if result[0]==2:
                        accepted_same += 1
                    if r == best[0] and accepted_same > 2:
                        break
                    T = 0.9*T
                    if verbose:
                        print r
                        if result[0] == 2:
                            print '*',
                        print median
                else: break
            return first, best

    if __name__ == "__main__":
        print 'Problem: simple network'
        a = network2distancematrix('../data/network-links', True)
        allpairs(a)
        result = simulated_annealing(a, 2, verbose=True)
        print result[0][0], result[1][0]

        print 'Problem: pmed1 in OR-lib'
        a = network2distancematrix('../data/orlib/pmed1.orlib',
                                   False)
        allpairs(a)
        result = simulated_annealing(a, 5, verbose=True)
        print result[0][0], result[1][0]
```

The program in Listing 12.6 includes the test for the two problems, the simple network and the first *p*-median problem, in the OR-Library, as was the case with the vertex exchange problem. We show the results of a sample run of the program below to illustrate some interesting features of the algorithm. First, we can see that optimal solutions to both problems are found. We will test how often this happens later. Second, it is important to note how (actually when) the optimal solution appears in both case. In our Python code, we print an asterisk (*) if the accepted solution is worse than the current solution. For example, in the run for the simple network problem, the optimal solution (objective function value 40) appears *after* the algorithm has accepted a worse solution (objective function value 45). This indicates that, at least for this case, the worse solution helped the algorithm find the optimal solution immediately after taking a step back. The same trend can be observed for the pmed1 problem, where three worse solutions are accepted before the optimal solution of 5,819 is found. Third, it is clear that the the best solutions are visited several times before the problem terminates. It is obvious here that our tricks worked in successfully avoiding unnecessary runs, which also saves computing time.

```
Problem: simple network
[52, [2, 1]]
43    [5, 1]
45 *  [5, 3]
40    [5, 0]
44 *  [4, 0]
40    [5, 0]
43 *  [5, 1]
45 *  [5, 3]
43    [5, 1]
52 40
Problem: pmed1 in OR-lib
[9462, [11, 13, 12, 81, 38]]
7175    [11, 3, 12, 81, 38]
6589    [11, 3, 12, 34, 38]
6144    [11, 3, 12, 34, 90]
5897    [98, 3, 12, 34, 90]
5907 *  [24, 3, 12, 34, 90]
5920 *  [24, 2, 12, 34, 90]
5899    [24, 65, 12, 34, 90]
5905 *  [24, 65, 41, 34, 90]
5856    [24, 65, 41, 6, 90]
5827    [24, 64, 41, 6, 90]
5821    [98, 64, 41, 6, 90]
5819    [98, 64, 12, 6, 90]
5843 *  [25, 64, 12, 6, 90]
5821    [24, 64, 12, 6, 90]
5819    [98, 64, 12, 6, 90]
5821 *  [24, 64, 12, 6, 90]
5827 *  [24, 64, 41, 6, 90]
5821    [24, 64, 12, 6, 90]
9462 5819
```

Now let us further examine how the acceptance probability changes during the search process. The probability is dependent on the temperature and the change in the objective function value. We write the following short program to examine the dynamics of the temperature and the acceptance probability based on three different increase values of the objective function:

```
from simulated_annealing import acceptable
T = 100.0
for i in range(100):
    print T, acceptable(1, T)[1], \
        acceptable(5, T)[1], \
        acceptable(10, T)[1]
    T = T*0.9
```

The results are plotted in Figure 12.1. The temperature curve shows a rapid decrease at the beginning of the procedure. The probabilities for the three objective function value changes showed a different trend as the biggest falls in probability occur between iterations 20 and 50. At the same iteration, a small delta value will yield a significantly higher probability than a large delta value.

We now test all 40 p-median problems in the OR-Library using the simulated annealing algorithm with the code in Listing 12.7.

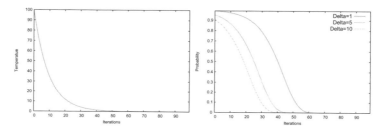

Figure 12.1 Temperature (left) and acceptance probability (right) used in the simulated annealing algorithm

Listing 12.7: Using the simulated annealing algorithm to test the 40 p-median problems in OR-Library (test_orlib_pmed_simann.py).

```
 1  import time
 2  from test_orlib_pmed import load_distance_matrix
 3  from simulated_annealing import *
 4  from test_orlib_pmed import *
 5
 6  def testall():
 7      for i in range(len(N)):
 8          fn = format("../data/orlib/pmed%d.distmatrix"%(i+1))
 9          dist = load_distance_matrix(fn)
10          t1 = time.time()
11          result = simulated_annealing(dist, pp[i])
12          t2 = time.time()
13          print format("pmed%d:"%(i+1)), N[i], pp[i],\
14              result[1][0], known[i], t2-t1,
15          if result[1][0]==known[i]: print "(*)"
16          else: print
17
```

```
18  def testpmed(i, numtest=20, neighbormethod=0):
19      fn = format("../data/orlib/pmed%d.distmatrix"%(i+1))
20      dist = load_distance_matrix(fn)
21      t1 = time.time()
22      sumdiff = 0
23      nummiss = 0
24      for j in range(numtest):
25          result = simulated_annealing(dist, pp[i],\
26              neighbormethod=neighbormethod)
27          print result[1][0],
28          if result[1][0]==known[i]:
29              print "(*)"
30          else:
31              print
32              sumdiff += result[1][0] - known[i]
33              nummiss += 1
34      t2 = time.time()
35      print "Time:", (t2-t1)/numtest
36      print "Found:", numtest-nummiss
37      if nummiss:
38          print "Avg diff:", float(sumdiff)/nummiss
39
40  if __name__ == "__main__":
41      testall()
42      testpmed(16)
43      testpmed(16, neighbormethod=1)
```

Here, we found optimal solutions for 15 out of the 40 problems, which is similar to the results for the vertex exchange algorithm which got 14 out of 40 for a single try. For those problems to which we did not find the optimal solutions, though, it appears the distances are larger than those in the vertex exchange results. For example, for problems pmed37 and pmed40, the simulated annealing reported the best solution with objective function value 5,080 and 5,143 respectively, while the vertex exchange algorithm had 5,067 and 5,134 respectively.

```
pmed1:  100  5 5819 5819 0.0839600563049 (*)
pmed2:  100 10 4093 4093 0.287177085876 (*)
pmed3:  100 10 4270 4250 0.191607952118
pmed4:  100 20 3098 3034 0.457482099533
pmed5:  100 33 1358 1355 0.779225826263
pmed6:  200  5 7824 7824 0.80565905571 (*)
pmed7:  200 10 5631 5631 0.725255012512 (*)
pmed8:  200 20 4445 4445 3.69324994087 (*)
pmed9:  200 40 2754 2734 9.48669195175
pmed10: 200 67 1265 1255 13.189332962
pmed11: 300  5 7696 7696 1.77202391624 (*)
pmed12: 300 10 6634 6634 4.31804585457 (*)
pmed13: 300 30 4374 4374 14.6342201233 (*)
```

```
pmed14:  300  60  2972  2968  34.3586940765
pmed15:  300 100  1734  1729  63.2703919411
pmed16:  400   6  7823  8162   1.8658220768
pmed17:  400  10  6999  6999   8.24238801003 (*)
pmed18:  400  40  4852  4809  45.7127890587
pmed19:  400  80  2870  2845  92.6540749073
pmed20:  400 133  1798  1789 164.285640001
pmed21:  500   5  9138  9138   3.01183795929 (*)
pmed22:  500  10  8579  8579  13.4972910881 (*)
pmed23:  500  60  4219  4619 126.247814894
pmed24:  500 100  2978  2961 262.334873199
pmed25:  500 167  1845  1828 508.916893959
pmed26:  600   5  9917  9917   7.942497015 (*)
pmed27:  600  10  8310  8307  15.1935739517
pmed28:  600  60  4524  4498 248.925246
pmed29:  600 120  3048  3033 497.998163939
pmed30:  600 200  2017  1989 977.196017981
pmed31:  700   5 10087 10086   7.5910961628
pmed32:  700  10  9301  9297  14.325248003
pmed33:  700  70  4727  4700 378.281183958
pmed34:  700 140  3020  3013 1355.75469995
pmed35:  800   5 10400 10400  11.208812952 (*)
pmed36:  800  10  9965  9934  22.7055418491
pmed37:  800  80  5080  5057 734.290642023
pmed38:  900   5 11060 11060  14.7149581909 (*)
pmed39:  900  10  9423  9423  35.8160691261 (*)
pmed40:  900  90  5143  5128 1251.52289701
```

It seems that the performance of simulated annealing is inferior to that of the vertex exchange algorithm. But is this a coincidence? This is hard to conclude. We can go further and examine a specific p-median problem. Let us first try pmed17 in the OR-Library to test the effectiveness of the simulated annealing algorithm. We ran this 20 times, with the following result:

```
7010
7003
7003
7033
6999 (*)
7003
6999 (*)
7014
7010
7010
6999 (*)
7010
6999 (*)
6999 (*)
```

```
7003
7025
7003
6999 (*)
7010
6999 (*)
Time: 4.70761389732
Found: 7
Avg diff: 11.5384615385
```

For seven out of 20 runs on the same problem, we were able to find the optimal solution. Though this is smaller than the result from the vertex exchange algorithm, the average difference from the optimal objective value is smaller, indicating that the two heuristics exhibit very similar effectiveness in finding optimal solutions, at least from the evidence of our tests so far.

All the above tests of simulated annealing use the `bestGeoNeighbor` function to generate neighboring solutions. Let us now take a brief look at the two different approaches to creating neighbors for a solution. The result listed below shows that the random method does not return any optimal solutions in 20 runs (only a portion of the results is shown). For this reason, we will only use the `bestGeoNeighbor` function for all the remaining tests.

```
8899
8628
...
8540
8555
0.323373091221
```

Both the vertex exchange and simulated annealing algorithms exhibited convergence toward an optimal or near-optimal solution. Figure 12.2 shows the overall trend of both algorithms from 20 runs on the same problem. Since the two algorithms are designed based on similar mechanisms of vertex replacement, a similar trend should be expected. However, we can also see that the vertex exchange algorithm was able to converge in

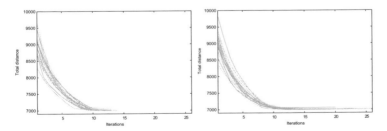

Figure 12.2 Convergence of the vertex exchange (left) and simulated annealing (right) algorithms on pmed17

relatively fewer iterations. It would be reasonable to attribute this quick convergence to the fact that vertex exchange exhaustively considers all possible replacements of all directly linked vertices (i.e., only those vertices that are directly linked to the selected vertices are considered for a replacement) in the current solution at each iteration, while the simulated annealing method only effectively considers a subset of the replacements, both between directly and non-directly linked vertices.

The comparison of convergence of the two algorithms also leads to a stunning difference in the computing time: simulated annealing as developed here takes a lot more time than its vertex exchange counterpart. It is obvious that the computing for simulated annealing can be controlled by changing the distance threshold. At the level of our threshold, vertices that are not directly connected with the selected ones will also be considered. Overall, we are considering more replacements in simulated annealing, which contributes to a significant increase in computing time. This also means the simulated annealing can possibly be "reduced" to the vertex exchange method if we require only direct links be considered for each iteration. In other words, the implementation of simulated annealing in this chapter can be considered as a general case of vertex exchange.

We now run a test on pmed1. The vertex exchange algorithm found the optimal solution for all 20 runs (in an average time of 0.058 seconds). The result for the simulated annealing algorithm is as follows:

```
5868
5868
5868
5868
5868
5819 (*)
5868
5819 (*)
5868
5819 (*)
5819 (*)
5821
5819 (*)
5819 (*)
5819 (*)
5819 (*)
5868
5868
5868
5819 (*)
Time: 0.101140642166
Found: 9
Avg diff: 44.7272727273
```

This prompts the question: is the simulated annealing algorithm less stable than the vertex exchange algorithm? One thing is clear: the amount of testing we have done so far does not allow us to draw clear conclusions. We will stop short here. The exercises section has more discussion on this issue.

12.4 Notes

A huge body of literature has been devoted to facility location and location-allocation problems (Daskin, 1995; Drezner, 1995). This is the cornerstone of what we refer to as spatial optimization in this book. The algorithms introduced in this and the previous chapters only represent a tiny portion of the area. The word "heuristic" is related to the Greek root of *eureka*, meaning "I have found it." When Archimedes discovered how he could measure the volume of an irregularly shaped object, he ran naked out into the street, crying "Eureka!" In our discussion of search algorithms, a heuristic is a method that can be used to find solutions to an optimization problem.

Many of the methods have been derived from a much wider spectrum of scientific research, especially from the computer science literature. Other than the simulated annealing (Kirkpatrick et al., 1983; Cerny, 1985) covered in this chapter, the metaheuristic methods, for example, include methods such as genetic algorithms (Goldberg, 1989), tabu search (Gendreau et al., 1997), ant colony systems (Dorigo and Gambardella, 1997), and more (Michalewicz and Fogel, 2000). Genetic algorithms belong to a family of metaheuristic methods called evolutionary algorithms. For the simulated annealing method there are other variations such as using a threshold instead of temperature (Dueck and Scheuer, 1990). These algorithms as a whole have been investigated intensively in various applications of spatial optimization problems (Krzanowski and Raper, 2001; Xiao et al., 2002, 2007; Xiao, 2006, 2008).

The vertex substitution method was developed by Teitz and Bart (1968). Even though the idea behind this heuristic is simple, many researchers have demonstrated that it is very effective in finding at least near-optimal solutions to the p-median problem (Rosing et al., 1979; Rosing, 1997; Xiao, 2012). There have been many developments since the vertex exchange method appeared (Densham and Rushton, 1992; Rosing and Hodgson, 2002; Resende and Werneck, 2004). All of these have been tested using benchmark problems of different kinds, including those in the OR-Library (Beasley, 1985). The original implementation of the algorithm, however, is not efficient. The `findout` algorithm used in vertex exchange was first developed by Whitaker (1983) and has brought about significant improvements in overall performance. It has been used in many other algorithms as well (Hansen and Mladenović, 1997; Resende and Werneck, 2003, 2004).

Since there are so many methods to choose from, at some point one must wonder which is the best. The "no free lunch" theorem (Wolpert and Macready, 1997) cautions us not to draw conclusions beyond what is tested. We have been careful not to draw conclusions about the two algorithms for the p-median problem. The fact is, even though we have tested intensively using a few problems, there are always more problems out there and we may have no clue how our heuristics will perform under new circumstances.

One way to address the variety of algorithms is to combine them into hybrid algorithms. The multi-start algorithm (Resende and Werneck, 2004) is a hybrid algorithm derived from components of many useful algorithms. It is also possible to treat each existing algorithm as a software agent, and we can allow cooperation between different agents so that they solve the problem together as a team (Xiao, 2012). These software agents can exchange the solutions using different mechanisms. They can also be distributed to different computers on the web to utilize the distributed computing environment so as to speed up the otherwise very time-consuming hybrid processes.

12.5 Exercises

1. We have designed a greedy algorithm for the p-median problem using the addition principle. In Section 12.1 we also mentioned that a deletion principle can be used to construct a feasible solution to a p-median problem. Implement a greedy algorithm using deletion and test it on the problems used in this chapter. Should we expect significantly different results from the greedy algorithm using addition?

2. Design and implement a greedy algorithm to solve the discrete p-center problem described in the exercises in Chapter 11.

3. Design and implement a greedy algorithm to solve the set covering problem described in the exercises in Chapter 11.

4. The 1-center problem discussed before is a discrete location problem that has a finite set of demand points, but the solution is in a continuous space. We can solve it exactly using one of the methods introduced. When there is a need to choose multiple centers, the problem becomes more difficult. Because the centers are continuous, we cannot write a mixed integer program that can only deal with discrete choices. This is the continuous p-center problem (where of course $p > 1$). Modify the k-means algorithm introduced in Chapter 9 so that it can be used as a heuristic method to solve the continuous p-center problem.

5. Test the greedy algorithm for all 40 p-median problems in the OR-Library. Report the time taken by the greedy algorithm for each problem.

6. A very important issue in designing heuristic methods is how to guide the search algorithm out of local optima. Compare and contrast the strategies in the three algorithms for the p-median problem discussed in this chapter. Consider your own strategies and see how effective they are. Try a few strategies to make your arguments about this issue.

7. Are there p-median problems that are bound to be easy? In other words, is it possible to design a network with a certain structure so that a search algorithm can always find the optimal solution regardless of the initial solution? On the other hand, we might also ask if there are networks that are bound to be difficult because there are some structures in the network that are designed to mislead the search process. Would any of these features be algorithm-specific, meaning they can fool some algorithms but not others? Thinking along these lines, design a toy network that exhibits some of these features.

8. We have used the 40 p-median problems from the OR-Library where the optimal solutions are known. What happens if we want to test more problems? Try to design a few problems of your own, with the optimal solutions known. You can try to use mixed integer problems to solve these problems.

Postscript

Donkey: Are we there yet?
Shrek: No.
Donkey: Are we there yet?
Fiona: Not yet.
Donkey: Are we there yet?
Fiona: No.
Donkey: Are we there yet?
Shrek: No!
Donkey: Are we there yet?
Shrek: Yes.
Donkey: Really?
Shrek: No!!

Shrek 2 (2004)

We started this book with an introduction to spatial data and some basic geometric algorithms. From there we moved on to discuss indexing methods that can be used to expedite the handling of spatial data. Then we explored a number of applications in spatial analysis. In each chapter, we focused on theory, methods, and practice with code. We used examples to show, in a humble way, that by coding we gain the freedom to do things without relying on large software packages. But are we there yet?

Each chapter in its own right could be extended into a book or even more than one book. We have covered some of the most important topics, concentrating our discussion around how the concepts work and how to write a simple program in Python to implement the methods. We discussed the depth of each area in the notes to each chapter, but only to a limited extent. More efforts, collective ones, will be needed to embrace the breadth and depth of the field involving geospatial data. Also, we can certainly do more with the code included. For example, each of the programs is used individually in this book, for a good reason – we need to introduce them individually for each algorithm when we go over the computational processes. Now that we have seen all the programs, we can compile them together into a Python package that can be used more easily. However, we should point out that the programs in this book are coded in a minimalist fashion so as to put the algorithms under the spotlight. While all the programs *work*, they may not work under all conditions. We do not check the type or validity of variables unless it is absolutely necessary to do so. For example, we do not make sure to only insert points (as in the `Point` class) into a *k*-D tree. These may become "bugs" that may cause a program to crash unexpectedly. Some may consider this to be a limitation of a weakly typed programming language like Python. However, it also gives us the convenience to think about the computational process from a more intuitive perspective by not going into excessive details.

Between here and there, then, there is the vast space of the discipline of geographic information science and spatial analysis, and rigorous software engineering processes. But none of that detracts from the true value of the chapters in this book: they are part of the process that prepares us to gain the freedom to handle spatial data.

Appendix A

Python: A Primer

Beautiful is better than ugly.
Explicit is better than implicit.
Simple is better than complex.
Complex is better than complicated.
Flat is better than nested.
Sparse is better than dense.
Readability counts.

Tim Peters, *The Zen of Python*

This appendix is mainly for those readers who may not be familiar with some of the Python features used in this book. It can be treated as a quick way to get started with Python but is not meant to be a thorough introduction to the language. For a comprehensive understanding of Python, it would be better to use other tutorials. There are excellent tutorials out there and many of them are free. For example, the official Python tutorial[1] covers a wide range of topics and provides many very effective examples that can be easily extended to more complicated cases. Other tutorials, such as *How to Think Like a Computer Scientist*,[2] may provide more specialized content. Nevertheless, here we will look at Python topics that are closely related to many of the algorithms introduced in this book. Our aim is to keep things simple as more detailed information is easily available on the Web or in other publications.

    Python is a cross-platform programming language. In this book we will assume a successful installation of Python and many of the modules that come with the installation. We only use the command line of the Python shell. We will introduce Python using some basic commands, before moving on to discuss some more specific topics. It should be noted that all Python commands are entered after the prompt `>>>`. The version of Python used to prepare this appendix is 2.7.6.

---

[1] `https://docs.python.org/2.7/tutorial/`

[2] `http://openbookproject.net/thinkcs/python/english3e/index.html`

We can use Python as a powerful calculator. However, we must be careful about the type of each item. A value of 2 will be treated as an integer; the calculation of 2/3 will be treated as an operation between two integers and the result will be an integer (hence 0) in the output. To ensure a floating point value result we must use a floating point value in at least one of the numbers in the operation (e.g., 2.0/3). We can also use the function float to specify the type of number.

```
>>> 10 * 10
100
>>> 2/3
0
>>> 2.0/3
0.6666666666666666
>>> 2/3.0
0.6666666666666666
>>> float(2)/3
0.6666666666666666
>>> 2/float(3)
0.6666666666666666
```

Variables are important in Python (as in all programming languages). However, variables in Python do not have types. The type of a variable will be determined by Python at run time. In the following example, we assign an integer to b first and then we change it to a string of text (enclosed in a pair of double quotes or single quotes). When we want to add a string to an integer, Python reports an error. Also note that in Python we use the symbol # to indicate comments and everything after the comment symbol in the same line will not be considered by Python for any operation.

```
>>> a = 12              # a variable with integer value
>>> b = 10              # another variable
>>> a + b               # adding two variables
22
>>> b = 'mystring'      # now variable b is a string
>>> a + b
Traceback (most recent call last):
  File "<stdin>", line 1, in <module>
TypeError: unsupported operand type(s) for +: 'int' and 'str'
```

If both variables are strings, we can definitely add them up. Note that a string can be indicated using either single or double quotes, as long as they match.

```
>>> a = "astring"
>>> a + ' ' + b
'astring mystring'
```

Strings can be duplicated using the "*" operation.

```
>>> a * 5
'astringastringastringastringastring'
```

```
>>> (a + ' ') * 5
'astring astring astring astring astring '
```

We can also combine variable values with strings to format powerful and meaningful expressions using the built-in `format` method that comes with each string. The first part of the command is a partial string where the braces are used as placeholders whose values will be filled in by the variables in the parentheses. The variables are indexed as 0, 1, and 2, and these indices are used in the partial string.

```
>>> n1, n2, n3 = 10, 1, 2
>>> 'I have {0} apples, {1} pears, and {2} others'.\
... format(n1, n2, n3)
'I have 10 apples, 1 pears, and 2 others'
```

The first line in the above example represents another useful feature of Python called simultaneous assignment that allows us to assign multiple values at once. Because of this feature, functions in Python can return multiple values at the same time.

```
>>> a,b = 3,4
>>> a,b
(3, 4)
>>> a,b = b,a
>>> a,b
(4, 3)
```

Another way to format a string is to use the `%` operation. This is a little more confusing but can be useful. Using `%` symbols within quotes indicates placeholders for values; such `%` symbols are followed by a character to indicate the type of variable or value. Here the letters `d` `f` and `s` denote integer, floating point, and string types, respectively. For a floating point value, we can also use `%.2f` to make sure only two decimal points are shown. A `%` symbol after an end quote indicates the variables or values that will be used to replace the placeholders in the quotes.

```
>>> n1, n2, n3 = 10, 1, 2
>>> "n1 = %d, n2 = %d, n3 = %d"%(n1, n2, n3)
'n1 = 10, n2 = 1, n3 = 2'
>>> name = "Dr. P"
>>> age = 40
>>> height = 6.1
>>> "%s is %d years old and %.2f feet tall"%(name,age,height)
'Dr. P is 40 years old and 6.10 feet tall'
```

Finally, the `help` function can be used to get useful explanations and examples of Python functions. Quotes must be used for Python symbols in the this function.

```
>>> help(format)
>>> help("<")
>>> help("==")
>>> help("%")
```

A.1 List comprehension

A list in Python can be used to store multiple things of any data type, and is the most powerful data structure in Python. In the following example, `a` is a list that contains six different values. The function `len` can be used to return the length of the list. Each item in a list is referred to using an index that starts from 0.

```
>>> a = ['str1', 10, 'str2', 'str3', 10, 5]
>>> a
['str1', 10, 'str2', 'str3', 10, 5]
>>> len(a)
6
>>> a[0] 'str1'
>>> a[0] + ' ' + a[2]
'str1 str2'
>>> a[1] + a[5]
15
```

When talking about lists, a very useful tool is the `range` function that returns a list of numbers. When only one argument (e.g., 10) is entered, it is the stop condition and the function will yield a list of integers from 0 to the integer one less than the input.

```
>>> range(9)
[0, 1, 2, 3, 4, 5, 6, 7, 8]
>>> range(2, 9)
[2, 3, 4, 5, 6, 7, 8]
>>> range(2, 9, 3)
[2, 5, 8]
```

Creating a list is called list comprehension in Python. There are many ways to create lists. Because a list contains a set of items, a traditional way to fill in values in these items is through a `for` loop. We first create an empty list (`[ ]`) and then keep adding new items to the list using the `append` function. The following example generates a list of 10 values, each being the *i*th power of 2 for *i* ranging from 0 to 9. (Note `**` in the following code is the operator for "power of" in Python.)

```
>>> exps = []
>>> for i in range(10):
...     exps.append(2**i)
...
>>> exps
[1, 2, 4, 8, 16, 32, 64, 128, 256, 512]
```

Before we move on to list comprehension, let us focus on a very important rule in Python syntax: indentation. Instead of using brackets or other symbols to indicate a code block, Python uses indentation. In the above example, the code that is logically within the `for` loop must be written with some leading white space (four spaces in this case). Before the indentation starts, the line above must end with a colon (:). This rule applies whenever a logical group of lines must be identified, as we will often see in this appendix and in the main text of the book.

The above example can be replaced using the list comprehension method in Python with one line of code (shown below) that returns a list with exactly the same contents. We also compare the two lists using the logical expression == that only returns `True` if elements in both lists are exactly the same. Each list comprehension contains at least two kinds of operations: assignment and enumeration. In our example below, the assignment (2**i) is done over an enumeration (using the `for` statement) of each of the `i` values from a list of 10 integers generated by the `range` function.

```
>>> exps1 = [2**i for i in range(10)]
>>> exps1 == exps
True
```

We can also put conditions in list comprehension to create more complicated lists. The following example creates a list with only the members from `exps1` that can be evenly divided by 32. Here, the symbol % is the modulo operator that returns the remainder from a division operation of the first parameter by the second (e.g., 5%3 returns 2).

```
>>> [x for x in exps1 if x%32==0]
```

We can actually plug in the expression for `exps1` to make a nested list comprehension.

```
>>> [x for x in [2**i for i in range(10)] if x%32==0]
```

In Section 2.8, we use the following two `for` loops to create a list of points for the parallels where each parallel contains 37 points ranging from −180 to 180 in steps of 10.

```
>>> points = []
>>> linenum = 0
>>> for lat in range(-90, 91, 10):
...     for lon in range(-180, 181, 10):
...         points.append([linenum, lon, lat])
...     linenum += 1
...
```

This can be simplified using list comprehension into one line of code as follows.

```
>>> points = [ [(lat+90)/10, lat, lon]
...            for lat in range(-90, 91, 10)
...            for lon in range(-180, 181, 10) ]
```

The *X* and *Y* coordinates of the points in the 17th parallel (80°N) can be retrieved using another list comprehension.

```
>>> [[p[1], p[2]] for p in points if p[0]==17]
```

A.2 Functions, modules, and recursive functions

As in any programming language, we can group a collection of lines of code into a function that can be used repeatedly. We use the Python keyword `def` to start defining a function. Immediately after the keyword is the name of the function followed by a set of input arguments in a pair of parentheses. The line defining a function ends with a colon. We use a simple example below to demonstrate how functions work.

Here we define a function that calculates the distance between two points with X and Y coordinates given by `x1` and `y1` for the first point and `x2` and `y2` for the second. We will need to use the `sqrt` function to compute the square root of the sum of squares. However, this function is not a built-in Python function. Instead, `sqrt` is available in a Python module called math that must be specifically imported (in the first line). All the lines in the function must be indented to indicate that they belong to the function. In this example, we do not show Python prompts because it is inconvenient to write multiple lines of code in the interactive Python shell where we cannot edit the previous lines. Instead, we can write the Python commands in a text editor and save them as a file. There are many good editors for us to choose from. My favorite is Emacs, but Notepad++ is also a popular and free option. Let us assume we save the following code in a file called `dist.py`. In this way, the code we write becomes part of a module and can be imported for use when needed.

```
import math
def distance(x1, y1, x2, y2):
    dx = x1-x2
    dy = y1-y2
    d = math.sqrt(dx*dx + dy*dy)
    print "my name is:", __name__
    return d
```

The following example shows how to use the code to calculate distance.

```
>>> from dist import *
>>> distance(1.5, 2.5, 3.5, 4.0)
my name is: dist
2.5
>>> print __name__
__main__
```

In this way, we have created a Python module called `dist`. In general, a Python module includes the code blocks that can be used somewhere else outside its original file. The above exercise also shows an important built-in variable in Python called `__name__`. This variable is assigned to the value of `__main__` in the process that was first executed by the Python interpreter, which in our case is the Python prompt. When a module is loaded, Python will use the module name to set the value of `__name__`.

The following is an example showing the calculation of the factorial of an integer using the formula $n! = n \times (n-1) \times (n-2) \times ... \times 3 \times 2 \times 1$. Here we simply use a list of integers ranging from 1 to n that repeatedly multiplies itself.

```
def factorial_i(n):
    f = 1
    for i in range(1,n+1):
        f = f*i
    return f
```

The calculation of *n!*, however, can be implemented in another manner where the function calls itself repeatedly. This is what we refer to as a recursive function.

```
def factorial_r(n):
    if n == 1:
        return 1
    return n * factorial_r(n-1)
```

Recursive functions are often simple and elegant to write. However, we must do this carefully to avoid endlessly calling the function. In general, each recursive function must have a base where the calling stops. In the above example, the last line returns by calling the function itself. If we do not control this, it will end up calling itself *ad infinitum*. Therefore, we always have a condition before we call the function itself. Here, if the value of n becomes 1, we will return a constant instead of another function call. This is the base case in this example. Also, a recursive function must have a way to move on so as to get closer to the base case. In our example, we make sure we will reach the base case by calling the function with a decreasing value of n. In the main text of this book there are plenty of examples of recursive functions, especially when we are dealing with trees where a recursive call makes it easy to go down the tree continuously until we reach the end of the tree (a leaf node). Using recursive functions, however, has a big disadvantage in terms of performance. For example, when n is 10 and we call function factorial_r(10-1), no answer is returned and we have to wait until n reaches 1 before we can go back and actually calculate all the values for n. This backtracking requires a lot of memory storage to hold the unsolved function calls, and by backtracking we will spend more time going through the loop.

A.3 Lambda functions and sorting

Sometimes, we need a function but do not need a real function with a name. This is an anonymous function and we use the Python construct lambda for this purpose. For example, below we define a function so that it returns True if an input is an even number or False otherwise. This is a quick way of defining a function without using the def keyword and all the syntax rules. It is quite simple: the variable before the colon is the input and everything after the colon will be calculated and returned.

```
>>> f = lambda x: x%2==0
>>> f(10)
True
>>> f(1)
False
```

This simple way of defining functions may seem unnecessary and confusing. Nevertheless, it provides great expressiveness and abstraction of our code. We demonstrate its use in the context of list sorting. A list in Python can be sorted using the built-in method `sort`.

```
>>> exps1 = [2**i for i in range(10)]
>>> exps1.sort(reverse=True)
>>> exps1
[512, 256, 128, 64, 32, 16, 8, 4, 2, 1]
```

However, this will not work when each element in the list has a more complicated structure. For example, if each element in the list is another list that contains the X and Y coordinates of a point, sorting that list directly will give results based on the first value in each element. How about if we want to sort it based on the second value in each element? Here is how the `lambda` function comes to our assistance. We utilize the `key` parameter of the `sort` function by specifying the second value in each element to be considered.

```
>>> points=[[10,1], [1,5], [1,2], [3,7], [10,8], [7,8], [9,8]]
>>> points.sort()
>>> points
[[1, 2], [1, 5], [3, 7], [7, 8], [9, 8], [10, 1], [10, 8]]
>>> points.sort(key=lambda points: points[1])
>>> points
[[10, 1], [1, 2], [1, 5], [3, 7], [7, 8], [9, 8], [10, 8]]
```

A.4 NumPy and Matplotlib

NumPy is a powerful Python module specifically developed to operate on arrays and matrices. We use some of the functions in NumPy extensively in Chapter 8 where interpolation methods are discussed. NumPy is closely tied to many of the features of Python. For example, we can easily convert a Python list into a NumPy array. Many of the NumPy functions are built to handle multiple inputs from an array directly. We demonstrate these features using the following simple example. Here, we apply the `sin` function in NumPy directly to a list called `angles`. It appears that NumPy implicitly converts the list into an array and computes the sine of each element, using these to form an array for output. The `sin` function that comes in the math module, however, does not work in this way because it requires a single float value to be the input.

```
>>> import numpy as np
>>> angles = [i*np.pi/10 for i in range(1, 10)]
>>> np.sin(angles)
array([0.30901699, 0.58778525, 0.80901699, 0.95105652, 1.,
       0.95105652, 0.80901699, 0.58778525, 0.30901699])
>>> import math
>>> math.sin(angles)
```

```
Traceback (most recent call last):
  File "<stdin>", line 1, in <module>
TypeError: a float is required
```

Matplotlib is a highly useful Python package designed for two-dimensional graphics. Despite its command line style, it is fairly straightforward and quick to use this package to make impressive plots and graphics. We show three examples here to demonstrate its use in three distinct areas that are related to our spatial data handling applications.

The Matplotlib package contains two modes, IPython and pylab. The pylab mode is the main interface to the functionality of Matplotlib and we import it first in the following example. Note that this way we actually also import NumPy as `np`. Then we create a list called `angles` to hold 60 angles for more than a cycle of 2π. The `plot` function creates a plot with the first two inputs as the X and Y coordinates. A number of parameters can be applied in the `plot` function to control the look of the plot. Nothing will be displayed until the `show` function is called. Figure A.1 shows the result of this example.

```
from pylab import *    # or: import matplotlib.pyplot as plt
angles = [i*pi/20 for i in range(60)]
plot(angles, np.sin(angles), color='grey')
plot(angles, np.cos(angles), linewidth=2, color='lightblue')
show()
```

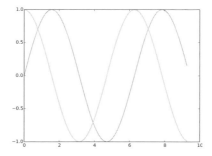

Figure A.1 Plotting using Matplotlib

The `plot` function can actually be used to plot line maps. We used this function in Chapter 2 to draw three map projections using the graticule and world coastlines. To plot polygons, we can use the `Polygon` function in Matplotlib.

The second example illustrates how to draw points in a scatter plot, which can be equivalent to a "point map." Here we create 100 points, 50 of which are scattered around the point (0,0) in a Cartesian coordinate system, and the other 50 around the point (10,10). Both X and Y coordinates follow a normal distribution with a standard deviation of 1. The NumPy function `normal` can be used to do that by specifying the mean, standard deviation, and number of samples to be created. We create the X and Y coordinates for each group separately and then combine them using the NumPy function `concatenate` for the X and Y coordinates, respectively. We mark the points in the first group around (0,0) in red and those in the second group around (10,10) in blue by creating a list called `col` with 100 elements, where the first 50 are `red` and the second 50 `blue`. Finally, the `scatter` function is used to create the plot, where we can also specify the colors for each point (`col`) and the transparency. A 0.5 level of transparency will allow us to show the effect when multiple points overlap. Figure A.2 shows the final result of this example. Note that we export the figure into a PDF, which is converted into the EPS format for publication separately because transparency in EPS is not natively supported in Matplotlib.

```
import matplotlib.pyplot as plt
import numpy as np
n = 100
X1 = np.random.normal(0, 1, n/2)
X2 = np.random.normal(10, 5, n/2)
Y1 = np.random.normal(0, 1, n/2)
Y2 = np.random.normal(10, 5, n/2)
X = np.concatenate((X1, X2))
Y = np.concatenate((Y1, Y2))
col = ['red' if i<n/2 else 'blue' for i in range(n)]
plt.scatter(X, Y, color=col, alpha=.5)
plt.axis('scaled')
plt.savefig('matplotlib-scatter.pdf',
            bbox_inches='tight',pad_inches=0)
plt.show()
```

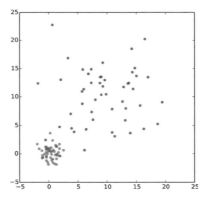

Figure A.2 Plotting points using Matplotlib

In our last example, we reuse the `distance` function we developed before to create a raster with two centers, one at the center of the lower-left quadrangle and the other at center of the upper-right. For each location in the raster, we find the nearest center to that location and set the value at the location using a biweight kernel (see Section 9.4). The function K is used to return such a value.

```
from dist import distance
def K(x, y, c1, c2, r):
    d1 = distance(x, y, c1[0], c1[1])
    d2 = distance(x, y, c2[0], c2[1])
```

```
        d, i = d1, 0
        if d2<d1:
             d,i = d2,1
        if d>r:
             v = 0
        else:
             u = float(d/r)
             v = 15.0/16.0 * (1 - u**2)**2
        return v
```

The following lines of code create a raster of 20 × 20 cells and the bandwidth of the kernel function is set to 10. We use list comprehension to create a list of 20 values from 1 to 20. The variables `c1` and `c2` are used to indicate the two centers. `Z` is the 2D list (a list of lists) where the value at each cell is generated using the function `K`. The Matplotlib function `imshow` is used to create a 2D image using values in `Z`. We use a color map of `Greys` to make a black-and-white map. By default, the `imshow` function will map the lowest value in the data to one end of the color map (white) and the highest to the other end (black). However, this will leave our map with very high contrast. Here, we force the function to use minimal and maximal values, −0.5 and 1.2, respectively, that are outside our data range (0 and 0.9375) so that the actual color used in the map will be in the middle range of the color map. The `origin` parameter in `imshow` specifies that the origin of the coordinates is in the lower-left corner of the image, which is consistent with our data. The `colorbar` function displays a legend and we set the total length of the bar to be 99% of the map height. The `xticks` and `yticks` functions take empty lists as input, which helps hide the ticks so that the rendering is more like a real map. Figure A.3 shows the result of this example.

```
from pylab import *
n = 20
r = n/2.0
c1 = [n/4, n/4]
c2 = [3*n/4, 3*n/4]
Z = [ [K(x, y, c1, c2, r) for x in range(1, n+1) ]
      for y in range(1, n+1)]
imshow(Z, cmap='Greys', vmin=-0.5, vmax=1.2, origin='lower')
colorbar(shrink=.99)
xticks([]), yticks([])
show()
```

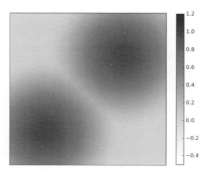

Figure A.3 Plotting a raster image using Matplotlib

A.5 Classes

We can group data and the functions that are closely associated with data into a class. In most Python terminology, data are called attributes and functions are called methods when they are encapsulated in a class. The simplest way to create a class is to use the `class` keyword in Python. We will create a class called `Shape` that stores a number of points. The first line declares the class. We define two methods in this class, both of them built-in methods. A built-in method in a Python class means it comes with every definition of the class and it does some specific thing. For example the `__init__` method is always called when an instance of the class is created. For this reason, we also call it the constructor of the class, and we as users can rewrite it with our own code. For example, here we require a `Shape` instance to be provided with a set of points when it is created. Each method in a Python class must include a variable called `self` as the first input argument. In this way, we can pass the class itself (along with its data and methods) to the method (function) for convenience. For example, in this `__init__` method, we can then assign the input variable values to the member attributes of the `Shape` class. Defining class attributes is straightforward: we simply write the names there with a `self.` prefix, where the dot means that whatever comes after the dot is a member of the class. The other built-in method is `__repr__` which defines how the instance of the class will be printed as text. Because of that, we will need to return a string. Here we simply convert the list of points into a string.

```
class Shape():
    def __init__(self, points, id=None, t=None):
        self.points = points
        self.id = id
        self.type = t
    def __repr__(self):
        return str(self.points)
```

To use the above class, we simply define a "variable" and assign it the class name as a function, with a set of input arguments. Such a variable is called an object of the class. Classes are a conceptualization or blueprint of the object. In other words, we cannot do anything with the class because it is just a definition, but we can create an object and operate on it.

```
>>> s = Shape([[1,1], [2,3], [3,7], [2,8]])
>>> print s
[[1, 1], [2, 3], [3, 7], [2, 8]]
```

There is not much we can do with the `Shape` object because an actual shape will be more complex than just a set of points. For example, we may have lines and polygons, and we will desire different behaviors from them. For this reason, we can extend an existing class into subclasses. This is called inheritance: we define subclasses that inherit features from their parent classes. We first take a look at an example of a subclass called `Polyline`. This time, we still use the `class` keyword, but we include the name of the parent class in the parentheses. In the constructor (`__init__`) of this class, we call the constructor of the parent class first to pass the necessary information that will be stored with the parent class. Beyond that, we introduce two new attributes to this new subclass, including `name` and `color`. Both of the values get a default value and we can make changes later. Now we also include a new method called `draw` where we first use the `Polygon` function in Matplotlib to create a line object and then we add it to the plot. Note that we can use the `Polygon` function to draw a line by specifying both parameters `closed` and `fill` to `False`. It will be necessary to import Matplotlib here.

```
import matplotlib.pyplot as plt
class Polyline(Shape):
    def __init__(self, points, id, name=None):
        Shape.__init__(self, points, id, "polyline")
        self.name = name
        self.color = 'grey'
    def draw(self):
        l = plt.Polygon(self.points, closed=False, fill=False,
                        edgecolor=self.color)
        plt.gca().add_line(l)
```

We then continue to define another class called `Polygon` below. For this subclass, we have two color properties, one for the fill (`fillcolor`) and one for the polygon outline (`edgecolor`). A `Polygon` class will have its own `draw` function due to the specific features of a polygon. We set the transparency level to 0.5 using the `alpha` parameter of the `Polygon` function.

```
class Polygon(Shape):
    def __init__(self, points, id, name=None):
        Shape.__init__(self, points, id, "polygon")
        self.name = name
        self.fillcolor = 'white'
        self.edgecolor = 'grey'
    def draw(self):
        poly = plt.Polygon(self.points, closed=True,
                           fill=True,
                           facecolor=self.fillcolor,
                           edgecolor=self.edgecolor,
                           alpha=0.5)
        plt.gca().add_line(poly)
```

Finally, we create another class called `Shapes` as a container to include and manage multiple shapes. The key feature here is a list called `shapes` and a function that adds different shapes to the list. Now we have a complete `draw` function that first calls all the shapes in the list to draw their geometric features (but not showing at this time). Then the `show` function in Matplotlib is called to finally draw the data.

```
class Shapes():
    def __init__(self):
        self.shapes = []
    def add_shape(self, shape):
        self.shapes.append(shape)
```

```
def draw(self):
    for s in self.shapes:
        s.draw()
    if len(self.shapes):
        plt.axis('scaled')
        plt.xticks([])
        plt.yticks([])
        plt.show()
```

We create three very simple shapes and draw the final map. Figure A.4 shows the result.

```
shapes = Shapes()
l1 = Polyline([[1,2], [5,3], [7,2]], 0)
l2 = Polyline([[6,2], [1,1], [5,5]], 1)
l2.color = 'red'
p1 = Polygon([[1,1], [2,3], [3,7], [2,8]], 3)

shapes.add_shape(l1)
shapes.add_shape(l2)
shapes.add_shape(p1)
shapes.draw()
```

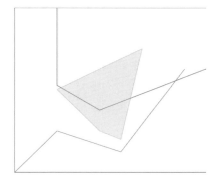

Figure A.4   Plotting shapes using Matplotlib

Appendix B

GDAL/OGR and PySAL

Many Python packages have been developed to handle geospatial data. In this appendix we introduce two sets of libraries for different purposes. GDAL (Geospatial Data Abstraction Library) is a powerful library that can be used to support various tasks in handling geospatial data sets. GDAL is designed to handle raster data and supports about 100 raster formats.[1] GDAL represents a major advance in free and open source GIS. As a component of GDAL, OGR supports vector data formats. OGR used to mean OpenGIS Simple Features Reference Implementation, but the OGR library does not fully comply with the OpenGIS Simple Features Specification of the Open GIS Consortium (OGC) and therefore was not approved by the OGC. As a consequence, OGR today technically does not stand for anything. In general, OGR supports many vector formats[2] such as shapefiles, personal geodatabases, ArcSDE, MapInfo, GRASS, TIGER/Line, SDTS, GML, and KML. It also supports many database connections, including MySQL, PostgreSQL, and Oracle Spatial. More information about GDAL and OGR can be found on their website.[3] GDAL/OGR was not initially developed as a Python library, but many programming languages support GDAL/OGR, including Python. In that sense, we are using a Python wrapper of the binary library of GDAL/OGR, which to some extent will have consequences on the performance of the code.

The second library is an open source software package called PySAL (Python Spatial Analysis Library) that has been used in a wide range of applications of spatial data.[4] Different from GDAL/OGR, PySAL is developed in Python and supports a wide range of spatial analysis applications. PySAL has its own core data structure that supports file

---

[1] A full list of the supported formats can be found at `http://www.gdal.org/formats_list.html`

[2] A full list of vector data formats supported under OGR can be found at `http://www.gdal.org/ogr/ogr_formats.html`

[3] `http://www.gdal.org`

[4] `http://pysal.org`

input and output. A key component of PySAL is spatial weights that are central in many spatial analysis methods.

In this appendix, we will go through some basic features to demonstrate how these powerful geospatial libraries can help us implement and understand many of the algorithms covered in this book. This appendix is mainly based on coding, with a lot of comments. Some of the lines included here are for instructional purposes and therefore may not be necessary in all cases. However, they should provide a good context in which to understand the functionality of GDAL/ORG and PySAL. We do not intend to explain the meaning of each and every line since it should be relatively straightforward to understand (especially with the extensive comments for most parts of the code). For more applications and details of the use of GDAL/OGR library, there is an online Python GDAL/ORG Cookbook.[5] PySAL has its own tutorials[6] that include details of how to use the library.

There is a great deal we can do with the power of GDAL/OGR and PySAL because these libraries give us a convenient way to convert geospatial data in any format into any data structures. Almost every algorithm introduced in this book can be plugged in here. We should take this opportunity to test and explore those algorithms using our own data.

Both GDAL/OGR and PySAL can be installed on different operating systems. We do not discuss the installation process here since the websites of these libraries provide detailed information.

B.1 OGR

To use OGR, we typically should start with the following lines, where we show how OGR works for a shapefile. In the following example, we assume there is a polygon shapefile called `uscnty48area.shp` existing in a folder called `data` in the upper level of the current working directory. To make it work for new data, we will need to have a polygon shapefile and make sure we know the path to that file. We focus on polygon shapefiles here. For other types of shapefiles (e.g., polyline), the processes are similar though the details will be different. We have an example of the use of a polyline shapefile (world coastlines) in Chapter 2. The last line in the following code should return `False`, indicating successful reading of the shapefile.

```
from osgeo import ogr                              # import OGR
drvName = "ESRI Shapefile"                         # driver/file type
driver = ogr.GetDriverByName(drvName)              # a shapefile driver
fname = '../data/uscnty48area.shp'                 # file name and path
vector = driver.Open(fname, 0)                     # 0-read, 1-writable
vector is None                                     # check if file is OK
```

---

[5]http://pcjericks.github.io/py-gdalogr-cookbook/

[6]http://www.pysal.org/users/tutorials/index.html

B.1.1 Attributes

The variable `vector` instantiated in the previous code lets us access the information stored in the shapefile. A general procedure is to go down the hierarchy of file > layer > feature to access different types of information. For a shapefile, one file only contains one layer, which in turn contains multiple features. Below are the lines of code for accessing the attributes associated with a feature.

```
layer = vector.GetLayer(0)               # shapefile uses 0
layer.GetFeatureCount()                  # count of features
feature = layer.GetFeature(0)            # the first feature
feature.items()                          # attributes feature
feature.GetFieldCount()                  # number of fields
fldnref = feature.GetFieldDefnRef(7)     # get field reference
fldn = fldnref.GetName()                 # get field name
feature.GetField(fldn)                   # get field val by name
feature.GetField(7)                      # get field val by id
```

We can write a short function (`getattr.py`) that returns a list of values in an attribute in a shapefile. We will use this function later in Section B.3.

```
from osgeo import ogr
import sys
def get_shp_attribute_by_name(shpfname, attrname):
    """
    Get attribute values by column name from a shapefile
    """
    driver = ogr.GetDriverByName("ESRI Shapefile")
    vector = driver.Open(shpfname, 0)
    layer = vector.GetLayer(0)
    f = layer.GetFeature(0)
    val = []
    for i in range(layer.GetFeatureCount()):
        f = layer.GetFeature(i)
        val.append(f.GetField(attrname))
return val
```

B.1.2 Geometry and coordinates

We will now see how to retrieve the geometry and coordinates from the shapefile we just loaded. We only show how to get points of one polygon, but it is of course possible to get points for all polygons using a loop, as we will show later.

```
layer.GetExtent()                        # extent of the layer
feature = layer.GetFeature(19)           # get a feature
geom = feature.GetGeometryRef()          # geom ref for feature
geom.GetEnvelope()                       # get envelope (extent)
geom.Centroid()                          # polygon centroid
```

```
geom.GetGeometryName()                # "POLYGON"
geom.GetPoint(0)                      # ERR 6 geom has rings
geom.GetGeometryCount()               # this geom has 1 ring
ring = geom.GetGeometryRef(0)         # get the first ring
ring.GetPointCount()                  # this has 31 points
p0 = ring.GetPoint(0)                 # first point on ring
pn = ring.GetPoint(30)                # last point on ring
p0 == pn                              # True: they are same
pn==ring.GetPoint(29)                 # False
```

B.1.3 Projecting points

We can transform a set of points in a geographic coordinate system (GCS) expressed in longitudes and latitudes into a projected coordinate system. Suppose we have a few locations from northern Cameroon that are recorded in WGS84 datum format, and we want to transform them into UTM zone 33N format. Here we will use the OSR (OGRSpatialReference) library, an OGR module for processing spatial reference systems, to transform between coordinate systems. OSR is based on the PROJ.4 Cartographic Projections Library.[7] The general flow of work is to start by defining the source coordinate system, which is what we call `gcs` in the following code. Then we define the target coordinate system, called `utmsr` below. There are different ways to define a projected coordinate system, and we choose the easiest way using the EPSG code that is uniquely assigned to UTM 33N. With these two coordinate systems, we can define a transformation using the OSR function `CoordinateTransformation`, and here we name the transformation `coordTransPcsUtm33`. The known coordinates we have for the four locations in Cameroon are stored in a list of lists where the string in each element list refers to a place name. We first need to convert these coordinates into a point geometry in OGR and then we can simply use the `Transform` function to do the coordinate system transformation.

```
from osgeo import ogr                           # import OGR
import osr                                      # import OSR
                                                #
gcs = osr.SpatialReference()                    # define a GCS
gcs.SetGeogCS("My GCS", "WGS_1984",
              "WGS84 Spheroid",
              osr.SRS_WGS84_SEMIMAJOR,
              osr.SRS_WGS84_INVFLATTENING,
              "Greenwich", 0.0, "degree")
utmsr = osr.SpatialReference()                  # empty spatial ref.
utmsr.ImportFromEPSG(32633)                     # EPSG code: UTM 33N
                                                # from gcs to utmsr
coordTransPcsUtm33 = osr.CoordinateTransformation(gcs, utmsr)
                                                # four places, in GCS
places = [ ["Mindif", 14.433333, 10.4], # Name, lon, lat
```

---

[7]http://trac.osgeo.org/proj/

```
            ["Pette", 14.5, 10.966667 ],
            ["Moulvoudaye", 14.85385, 10.40588],
            ["Diamare", 10.89510, 14.80438]]
point = ogr.Geometry(ogr.wkbPoint)        # empty point geometry
for p in places:
    point.AddPoint(p[1], p[2])            # X (lon), Y (lat)
    point.Transform(coordTransPcsUtm33)   # transform
    print p[0], point                     # print out the result
```

B.1.4 Projecting and writing geospatial data

Now we turn to our shapefile that is in a GCS and this time we will project it to the Albers equal-area conic projection that is commonly used for the conterminous United States. Our shapefile comes with a .prj file and we can read the spatial reference information directly from there. The general idea here is to find each point and transform it to a new coordinate system and then write it to a ring, which is in turn added into a polygon feature. We save the transformed data into a new shapefile.

A general polygon in a shapefile may have holes, where each hole and the outline polygon are called rings. However, many counties in our data also have islands in a river or lake and these islands are not holes. In this case, they are organized as multipolygons. As the name suggests, each multipolygon includes a set of polygons, and each of those polygons may still contain holes. To streamline the process of handling such data, we design a specific function `trans_rings` that retrieves every ring in the input geometry that must be a polygon, which must not be a multipolygon but can have holes. (Having holes or not does not affect how this function works because holes and the outline polygon are all treated as rings and a polygon will have at least one ring.) This function returns a new polygon that contains transformed points. This polygon is then added into the feature for the new shapefile. For a feature that is only a simple polygon (with or without holes), we call this function once and get a new polygon. For a multipolygon feature, we will need to call this function multiple times, each time for a polygon in the multipolygon. Every time a new polygon is added, we create a new feature and add the polygon into the feature.

```
import os                                      # import OS module
from osgeo import ogr                          # import OGR
import osr                                     # import OSR
                                               #
drvName = "ESRI Shapefile"                     # driver format
driver = ogr.GetDriverByName(drvName)          # shapefile driver
fname = '../data/uscnty48area.shp'             # input file name
vector = driver.Open(fname, 0)                 # open input file
layer = vector.GetLayer(0)                     # shapefiles use 0
ofname = 'uscnty48_proj.shp'                   # output file name
if os.path.exists(ofname):                     # if file exists
    driver.DeleteDataSource(ofname)            # we will delete it
                                               #
ovector = driver.CreateDataSource(ofname)      # output vector in
```

```
olayer = ovector.CreateLayer('POLYGONS',    # point shapefile
                          geom_type=ogr.wkbPolygon)
fieldDefn = ogr.FieldDefn('id',
                       ogr.OFTInteger)# integer field id
olayer.CreateField(fieldDefn)               # field in output file
featureDefn = olayer.GetLayerDefn()         # feature for later use

gcs = layer.GetSpatialRef()                 # geog coord system
pcs = osr.SpatialReference()                # empty proj coord sys
pcs.SetProjCS("Albers Conic Equal Area")    # conterminous USA
pcs.SetACEA(29.5, 45.5,                     # std parallels: 29.5, 45.5
           23, -96, 0, 0)                   # center meridian: 23N 96W
pcs.SetWellKnownGeogCS("NAD83")             # set datum (NAD83)
                                            # transform from gcs to pcs
coordTrans = osr.CoordinateTransformation(gcs, pcs)

pcs.MorphToESRI()                           # convert pcs to .prj
fprj = open(ofname.strip('.shp')+'.prj', 'w')
fprj.write(pcs.ExportToWkt())
fprj.close()

def trans_rings(geom):                      # return a polygon
    polygon = ogr.Geometry(
        ogr.wkbPolygon)                     # an empty polygon
    point = ogr.Geometry(ogr.wkbPoint) # a temp point
    for ring in geom:                       # loop all rings
        ring1 = ogr.Geometry(
            ogr.wkbLinearRing)              # create an empty ring
        for p in ring.GetPoints():          # loop points in ring
            point.AddPoint(p[0], p[1]) # make a point
            point.Transform(coordTrans)# transform GCS to PCS
            ring1.AddPoint(
            point.GetX(),
            point.GetY(),
            point.GetZ())                   # make the new ring
        polygon.AddGeometry(ring1)          # add ring to polygon
    return polygon

k = 0
for f in layer:                             # loop layer features
    geom = f.GetGeometryRef()               # get feature geometry
    ofeat = ogr.Feature(featureDefn)        # output feature
    ofeat.SetField('id', k)                 # set id of the feature
    geomtype = geom.GetGeometryName()
    if geomtype == "MULTIPOLYGON":
        for geom1 in geom:
            polygon = trans_rings(geom1)
            ofeat.SetGeometry(polygon) # set polygon to ofeat
            olayer.CreateFeature(ofeat)# attach ofeat to layer
    elif geomtype == "POLYGON":
        polygon = trans_rings(geom)
```

```
            ofeat.SetGeometry(polygon)      # same as above
            olayer.CreateFeature(ofeat)     # same as above
        else:
            pass                            # none polygon types
        k += 1

vector.Destroy()                            # close input shapefile
ovector.Destroy()                           # close output shp
```

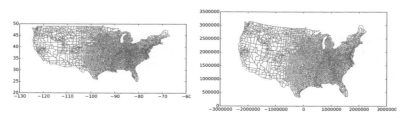

Figure B.1 Projecting spatial data. Left: original data in GCS. Right: data projected into Albers equal-area conic projection

The result is shown in Figure B.1. The two maps in this figure are drawn using Matplotlib using the following Python code (`draw_shp_polygons.py`).

```
from osgeo import ogr
import matplotlib.pyplot as plt
import sys

def plot_rings(geom):                       # plot
    ptsx = []
    ptsy = []
    for ring in geom:                       # loop all rings
        points = ring.GetPoints()           # get points in ring
        ptsx += [p[0] for p in points]
        ptsy += [p[1] for p in points]
    plt.plot(ptsx, ptsy, color='grey')

drvName = "ESRI Shapefile"
driver = ogr.GetDriverByName(drvName)       # a shapefile driver

driver = ogr.GetDriverByName("ESRI Shapefile")
if len(sys.argv) == 2:
    fname = sys.argv[1]                     # file name input
else:
    print "Usage:", sys.argv[0], "FILE.shp"
    sys.exit()

vector = driver.Open(fname, 0)              # open input file
layer = vector.GetLayer(0)                  # shapefiles use 0
```

```
for f in layer:
    geom = f.GetGeometryRef()           # geometry of feature
    geomtype = geom.GetGeometryName()
    if geomtype == "MULTIPOLYGON":
        for geom1 in geom:
            plot_rings(geom1)
    elif geomtype == "POLYGON":
        plot_rings(geom)
    else:
        pass                             # none polygon types
    f.Destroy()                          # remove input feature

vector.Destroy()                         # close the shapefile
plt.axis('scaled')
plt.savefig('us48prj.eps', bbox_inches='tight', pad_inches=0)
plt.show()
```

B.1.5 Adjacency matrix

Our final project using OGR is to create an adjacent matrix for a polygon shapefile. The procedure is simple: we continuously check each pair of polygons and test if they touch each other or one contains another. We use the OGR `Touches` and `Contains` functions to do so. To speed up the process, we design a function called `env_touch` to test if the bounding boxes (envelopes) of two polygons touch each other. We can write a function called `geom_touch` to test whether two OGR geometry objects touch each other by sharing a specified number of points (in parameter n0).

The `adjacency_matrix` function takes three input parameters: the path to the shapefile, the output format (M for matrix and L for list), and the number of shared points for polygons to be considered as adjacent (default is 1). In the test of the function, we use the US county shapefile and dump the final result in matrix form in a Python pickle file. In Python, pickle is a standard way of translating complicated objects so that we can store them in a text file and restore them for future use. The pickle file stored here is used in our main text (Section 9.2) for calculating Moran's *I*.

```
from osgeo import ogr
import numpy as np
import pickle

XMIN = 0    # envelope[0]: xmin
XMAX = 1    # envelope[1]: xmax
YMIN = 2    # envelope[2]: ymin
YMAX = 3    # envelope[3]: ymax

def env_touch(e1, e2):
    if e1[XMAX]<e1[XMIN] or e1[XMIN]>e2[XMAX] or\
       e1[YMAX]<e2[YMIN] or e1[YMIN]>e2[YMAX]:
        return False
    return True

def geom_share(g1, g2, n0):
    pts1 = []
```

```python
        if g1 is None or g2 is None:
            return False
        for g in g1:
if g.GetGeometryCount()>0:
    for gg in g:
        pts1.extend(gg.GetPoints())
    else:
        pts1.extend(g.GetPoints())
pts2 = []
for g in g2:
            if g.GetGeometryCount()>0:
                for gg in g:
                    pts2.extend(gg.GetPoints())
            else:
                pts2.extend(g.GetPoints())
    np = 0
    for p1 in pts1:
        for p2 in pts2:
            if p1==p2:
                np+=1
            if np>=n0:
                return True
    return False

def adjacency_matrix(shpfname, output="M",
                     num_shared_points=1):
    driver = ogr.GetDriverByName("ESRI Shapefile")
    vector = driver.Open(shpfname, 0)
    layer = vector.GetLayer(0)
    n = layer.GetFeatureCount()
    if output=="M":
        adj = np.array([[0]*n for x in range(n)])
    elif output=="L":
        adj = []
    else:
        return None
    for i in range(n):
        feature1 = layer.GetFeature(i)
        geom1 = feature1.GetGeometryRef()
        env1 = geom1.GetEnvelope()
        for j in range(i):
            feature2 = layer.GetFeature(j)
            geom2 = feature2.GetGeometryRef()
            env2 = geom2.GetEnvelope()
            if not env_touch(env1, env2):
                continue
            is_adj = False
            if geom1.Touches(geom2):
                if geom_share(geom1, geom2,
                              num_shared_points):
                    is_adj = True
            elif geom1.Contains(geom2):
                is_adj = True
```

```
                if is_adj:
                    if output=="M":
                        adj[i][j] = adj[j][i] = 1
                    elif output=="L":
                        adj.append([i, j])
                else:  # undefined
                    pass
    return adj

if __name__ == "__main__":
    shpfname = '../data/uscnty48area.shp'
    shpadj = adjacency_matrix(shpfname)
    pickle.dump(shpadj, open('uscnty48area.adj.pickle', 'w'))
```

B.2 GDAL

We now turn to GDAL to explore a few ways to access raster data. In this example we use the 2011 National Land Cover Database (NLCD) to demonstrate the use of GDAL. Similarly to OGR, we will need a driver to start using raster data, but in GDAL we also need to call the `Register` function before we can actually use it to open the raster file. Driver names (code) are available at http://www.gdal.org/formats_list.html. The NLCD we use here is stored in Erdas Imagine format (.img), and the GDAL code for that is `HFA`. The purpose here is to load the NLCD data into a NumPy array using the `ReadAsArray` function. Once the data are in an array format, which is an intuitive format for raster data, we can process them using many other methods such as the contagion metric (Section 9.4).

```
from osgeo import gdal                         # import GDAL
driver = gdal.GetDriverByName('HFA')           # Erdas Imagine driver
driver.Register()                              # register the driver
fname = "nlcd_2011_landcover_2011_edition_2014_03_31.img"
raster = gdal.Open(fname,
                   gdal.GA_ReadOnly)
raster is not None                             # file is loaded?
ncols = raster.RasterXSize                     # raster size in x
nrows = raster.RasterYSize                     # raster size in y
raster.RasterCount                             # number of bands
geotransform = raster.GetGeoTransform()        # spatial ref. info
ul_x = geotransform[0]                         # upper left x
x_res = geotransform[1]                        # west-east resolution
geotransform[2]                                # rotation on X
ul_y = geotransform[3]                         # upper left y
geotransform[4]                                # rotation on Y
y_res = geotransform[5]                        # north-south res.
band = raster.GetRasterBand(1)                 # get band (1-based)
x = 835000                                     # X (central Ohio)
y = 1900000                                    # Y (central Ohio)
```

```
Xoff = int((x - ul_x) / x_res)            # offset on X (110934)
Yoff = int((y - ul_y) / y_res)            # offset on Y (47000)
data1 = band.ReadAsArray(Xoff, Yoff,
                         1, 1)            # get 1x1 grid of data
data1[0, 0]                               # value at (0, 0)
data=band.ReadAsArray(0,0,ncols,nrows)    # make a numpy array
data[47000, 110934]                       # value at loc.
band=None                                 # release memory
data=None
```

B.3 PySAL

PySAL covers a very wide range of topics in spatial analysis and beyond (Rey and Anselin, 2010). Here we demonstrate how to use it to calculate Moran's *I* for an area data set. PySAL provides a lot of options to select the type of spatial weights. Here we use binary weights, meaning the weight between two polygons will be either 1 (adjacent) or 0 (non-adjacent). There are also a few different ways to compute the adjacency between polygons, and here we use the queen criterion, in which two polygons are adjacent if they share one or more points (the rook criterion, on the other hand, would require that they share an edge). Based on these choices, we will be able to compare the PySAL results with the code discussed in this book.

```
import pysal                              # import PySAL
import numpy as np                        # import NumPy
fname = '../data/franklin.shp'            # Franklin County data
w = pysal.queen_from_shapefile(fname)     # neighbors and weights
w.neighbors[0]                            # 1st polygon neighbors
sum([len(w.neighbors[i])
     for i in range(w.n)])                # # bi-directed edges
dbname = '../data/franklin.dbf'           # attribute table
db = pysal.open(dbname, 'r')              # get the attributes
db.header                                 # show attributes' names
y = np.array(db.by_col('Franklin_b'))     # use a column
mi = pysal.Moran(y, w,
                 transformation='B')      # Moran's I binary w
mi.I                                      # 0.31583739278406564
mi.EI                                     # expected I

from adjacency_matrix import *            # adjacency matrix code
from getattr import *
import sys
sys.path.append('../spatialanalysis')
from moransi2 import *                    # import moransi2
adj = adjacency_matrix(fname,             # compute adjacent
                       output="L")        #   list as output
len(adj)                                  # undirected edges
y1 = get_shp_attribute_by_name(
    fname, "Franklin_b")                  # get attributes
moransi2(y1, adj)                         # 0.3158373927840651
```

In Section 9.2, we introduced the calculation of Moran's I based on an adjacency matrix (or list), which is equivalent to the binary weights determined using the queen criterion. The above code also tests the result of using the Moran's I code in this book and the results are almost identical (readers who are interested in this example can examine the source code and explain what causes the very small difference between the two programs). To further validate the results, we can test the same data using the spdep package in R. To the precision reported in the R package, the result is 0.3158373928, which is consistent with the results obtained by the Python programs.

Appendix C

Code List

The majority of the programs in this book have a file name so that they can be imported as modules or run from the command line. Each program is typically run in its own directory, and whenever we need to use a module from another directory we append that directory to the system path using `sys.path.append` from the sys module. Suppose we are in directory `interpolation` and we need to import the `point.py` module from the `geom` directory. We can use the following three lines:

```
import sys
sys.path.append('../geom')
from point import *
```

In Table C.1 we list all the named files, in the order they appear in the text. In addition to these files, there are code blocks in the text that are not listed here because they are mainly meant for interactive commands. Data sets used as examples in this book are also listed in the table.

Table C.1 Code and data

Folder	Name	Section
geom	point.py	2.1
	spherical_distance.py	2.2
	point2line.py	2.3
	centroid.py	2.4
	sideplr.py	2.5
	linesegment.py	2.6
	intersection.py	2.7
	point_in_polygon.py	2.8
	point_in_polygon_winding.py	2.9

Folder	Name	Section
	neville.py	2.10
	transform1.py	2.11
	worldmap.py	2.12
	test_projection.py	2.13
	transform2.py	2.14
	line_seg_eventqueue.py	3.2
	line_seg_intersection.py	3.3
	test_line_seg_intersection.py	3.4
	overlay.py	3.5
	test_overlay_1.py	3.6
indexing	kdtree1.py	5.1
	kdtree2a.py	5.2
	kdtree2b.py	5.3
	kdtree3.py	5.4
	prkdtree1.py	5.5
	prkdtree2.py	5.6
	prkdtree3.py	5.7
	kdtree_performance.py	5.8
	quadtree1.py	6.1
	pointquadtree1.py	6.2
	pointquadtree2.py	6.3
	pointquadtree3.py	6.4
	xdcel.py	7.1
	extent.py	7.2
	pmquadtree.py	7.3
	pm1quadtree.py	7.4
	pm2quadtree.py	7.5
	pm3quadtree.py	7.6
	rtree1.py	7.7
	rtree2.py	7.8
	use_rtree.py	7.9
interpolation	idw.py	8.1
	read_data.py	8.2
	prepare_interpolation_data.py	8.3
	test_idw.py	8.4
	idw_cross_validation.py	8.5
	semivariance.py	8.6
	covariance.py	8.8
	fitsemivariance.py	8.9
	test_fitsemivariance.py	8.10
	okriging.py	8.11
	ordinary_kriging_test.py	8.12

Folder	Name	Section
	skriging.py	8.13
	simple_kriging_test.py	8.14
	interpolate_surface.py	8.15
	kriging_cross_validation.py	8.16
	midpoint2d.py	8.17
spatialanalysis	nnd.py	9.1
	nnd_monte_carlo.py	9.2
	oval_trees.py	9.3
	test_oval_trees_nnd.py	9.4
	kfunction.py	9.5
	test_oval_trees.py	9.6
	moransi.py	9.7
	moransi2.py	9.8
	test_moransi.py	9.9
	test_moransi_mc.py	9.10
	test_moransi_grid.py	9.11
	kmeans.py	9.12
	contagion.py	9.13
	test_contagion.py	9.14
	test_contagion_kernel.py	9.15
networks	network2listmatrix.py	10.2
	bfs.py	10.3
	dfs.py	10.4
	dijkstra.py	10.5
	gateway.py	10.6
	allpairdist.py	10.7
	allpairpaths.py	10.8
optimization	disc.py	11.1
	cp.py	11.2
	elzinga_hearn.py	11.3
	welzl.py	11.4
	test_1center.py	11.5
	pmed_lpformat.py	11.6
	pmed_greedy.py	12.1
	teitz_bart.py	12.3
	test_orlib_pmed.py	12.4
	simulated_annealing.py	12.6
	test_orlib_pmed_simann.py	12.7
gdal	getattr.py	B.1.1
	transform_cameroon_points.py	B.1.3

Folder	Name	Section
	transform_coordinate_systems.py	B.1.4
	draw_shp_polygons.py	B.1.4
	adjacency_matrix.py	B.1.5
	gdal1.py	B.2
	pysal1.py	B.3
data	uscnty48area.*	
	franklin.*	
	necoldem.dat	
	necoldem250.dat	
data/orlib	pmed*.orlib	
	pmed*.distmatrix	

References

Aluru, S. (2005). Quadtrees and octrees. In D. Metha and S. Sahni (eds), *Handbook of Data Structures and Applications*. Boca Raton, FL: Chapman & Hall/CRC.

Anselin, L. (1995). Local indicators of spatial association – LISA. *Geographical Analysis* 27(2), 93–115.

Bailey, T. C. and A. C. Gatrell (1995). *Interactive Spatial Data Analysis*. Harlow, UK: Pearson Education.

Bayer, R. and E. McCreight (1972). Organization and maintenance of large ordered indexes. *Acta Informatica 1*, 173–189.

Beasley, J. E. (1985). A note on solving large p-median problems. *European Journal of Operational Research 21*, 270–273.

Beckmann, N., H. Kriegel, R. Schneider, and B. Seeger (1990). The R*-tree: An efficient and robust method for points and rectangles. In *Proceedings of the ACM SIGMOD Conference on Management of Data*, Atlantic City, pp. 322–331. New York: Association for Computing Machinery.

Bentley, J. L. (1975). Multidimensional binary search trees used for associative searching. *Communications of the ACM 18*(9), 509.

Bentley, J. L. and T. A. Ottmann (1979). Algorithms for reporting and counting geometric intersections. *IEEE Transactions on Computers C 28*(9), 643–647.

Bentley, J. L., D. F. Stanat, and J. E. H. Williams (1977). The complexity of finding fixed-radius near neighbors. *Information Processing Letters 6*(6), 209–212.

Boots, B. N. and A. Getis (1988). *Point Pattern Analysis*. Newbury Park, CA: Sage.

Burrough, P. A. and R. A. McDonnell (1998). *Principles of Geographical Information Systems*. Oxford: Oxford University Press.

Carr, J. R. (1995). *Numerical Analysis for the Geological Sciences*. Englewood Cliffs, NJ: Prentice Hall.

Cerny, V. (1985). Thermodynamical approach to the traveling salesman problem: An efficient simulation algorithm. *Journal of Optimization Theory and Application 45*(1), 41–51.

Chakraborty, R. K. and P. K. Chaudhuri (1981). Note on geometrical solution for some minimax location problems. *Transportation Science 15*, 164–166.

Chan, T. M. (1994). A simple trapezoid sweep algorithm for reporting red/blue segment intersections. In *Canadian Conference on Computational Geometry*, Saskatoon, Saskatchewan, Canada, pp. 263–268.

Chrystal, G. (1885). On the problem to construct the minimum circle enclosing n given points in the plane. *Proceedings of Edinburgh Mathematical Society 3*, 30–33.

Clark, P. J. and F. C. Evans (1954). Distance to nearest neighbor as a measure of spatial relationships in populations. *Ecology 35*, 445–453.

Cliff, A. D. and J. K. Ord (1981). *Spatial Processes: Models and Applications*. London: Pion.

Cormen, T. H., C. E. Leiserson, R. L. Rivest, and C. Stein (2001). *Introduction to Algorithms* (2nd edn). Cambridge, MA: MIT Press.

Cressie, N. (1990). The origins of kriging. *Mathematical Geology 22*, 239–253.

Cressie, N. A. C. (1991). *Statistics for Spatial Data*. New York: John Wiley & Sons.

Daskin, M. S. (1995). *Network and Discrete Location: Models, Algorithms, and Applications*. New York: John Wiley & Sons.

de Berg, M., M. van Kreveld, M. Overmars, and O. Schwarzkopf (1998). *Computational Geometry: Algorithms and Applications* (2nd edn). Berlin: Springer.

Densham, P. J. and G. Rushton (1992). A more efficient heuristic for solving large p-median problems. *Papers in Regional Science 41*(3), 307–329.

Deza, M. M. and E. Deza (2010). *Encyclopedia of Distances* (2nd edn). Berlin: Springer.

Dijkstra, E. W. (1959). A note on two problems in connexion with graphs. *Numerische Mathematik 1*, 269–271.

Dorigo, M. and L. M. Gambardella (1997). Ant colony system: A cooperative learning approach to the traveling salesman problem. *IEEE Transactions on Evolutionary Computation 1*(1), 53–66.

Drezner, Z. (ed.) (1995). *Facility Location: A Survey of Applications and Methods*. New York: Springer.

Drezner, Z. and S. Shelah (1987). On the complexity of the Elzinga–Hearn algorithm for the 1-center problem. *Mathematics of Operations Research 12*(2), 255–261.

Dueck, G. and T. Scheuer (1990). Threshold accepting: A general purpose optimization algorithm appearing superior to simulated annealing. *Journal of Computational Physics 90*, 161–175.

Elzinga, J. and D. W. Hearn (1972). Geometrical solutions for some minimax location problems. *Transportation Science 6*, 379–394.

Finkel, R. and J. Bentley (1974). Quad trees: A data structure for retrieval on composite keys. *Acta Informatica 4*(1), 1–9.

Floyd, R. W. (1962). Algorithm 97: Shortest path. *Communications of the ACM 5*(6), 345.

Fotheringham, A. S., C. Brunsdon, and M. Charlton (2000). *Quantitative Geography: Perspectives on Spatial Data Analysis*. London: Sage.

Fournier, A., D. Fussell, and L. Carpenter (1982). Computer rendering of stochastic models. *Communication of the ACM 25*(6), 371–384.

Frank, A. U. (1987). Overlay processing in spatial information systems. In *Proceedings, Eighth International Symposium on Computer-Assisted Cartography (Auto-Carto 8)*, pp. 12–31. Baltimore, MD: ASPRS, ACSM.

Franklin, W. R., C. Narayanaswami, M. Kankanhalli, D. Sun, M.-C. Zhou, and P. Y. Wu (1989). Uniform grids: A technique for intersection detection on serial and parallel machines. In *Ninth International Symposium on Computer-Assisted Cartography (Auto-Carto 9)*, pp. 100–109. Baltimore, MD: ASPRS, ACSM.

Geary, R. C. (1954). The contiguity ratio and statistical mapping. *Incorporated Statistician 5*, 115–141.

Gendreau, M., G. Laporte, and F. Semet (1997). Solving an ambulance location model by tabu search. *Location Science 5*(2), 75–88.

Getis, A. and J. K. Ord (1992). The analysis of spatial association by use of distance statistics. *Geographical Analysis 24*(3), 189–206.

Goldberg, D. E. (1989). *Genetic Algorithms in Search, Optimization and Machine Learning*. Reading, MA: Addison-Wesley.

Goovaerts, P. (1997). *Geostatistics for Natural Resources Evaluation*. Oxford: Oxford University Press.

Guttman, A. (1984). R-trees: A dynamic index structure for spatial searching. In *Proceedings of the 1984 ACM SIGMOD International Conference on Management of Data*, SIGMOD 84, Boston, pp. 47–57. New York: ACM.

Haines, E. (1994). Point in polygon strategies. In P. S. Heckbert (ed.), *Graphics Gems IV*, pp. 24–46. San Francisco: Morgan Kaufmann.

Hansen, P. and N. Mladenović (1997). Variable neighborhood search for the *p*-median. *Location Science 5*(4), 207–226.

Healey, R. G., S. Sowers, B. M. Gittings, and M. J. Mineter (1998). *Parallel Processing Algorithms for GIS*. London: Taylor and Francis.

Huang, C.-W. and T.-Y. Shih (1997). On the complexity of point-in-polygon algorithms. *Computers & Geosciences 23*(1), 109–118.

Ipbuker, C. (2004). Numerical evaluation of the Robinson projection. *Cartography and Geographic Information Science 31*(2), 79–88.

Jordan, C. (1887). *Cours d'analyse de l'École polytechnique*. Paris: Gauthier-Villars.

Kamel, I. and C. Faloutsos (1994). Hilbert R-tree: An improved R-tree using fractals. In *Proceedings of the 20th International Conference on Very Large Data Bases*, pp. 500–509. San Francisco: Morgan Kaufmann.

Kirkpatrick, S., C. D. Gelatt, and M. P. Vecchi, Jr. (1983). Optimization by simulated annealing. *Science 220*, 671–680.

Krige, D. G. (1951). A statistical approach to some basic mine valuation problems on the Witwatersrand. *Journal of the Chemical and Metallurgical Mining Society of South Africa 52*, 119–139.

Krzanowski, R. M. and J. Raper (2001). *Spatial Evolutionary Modeling*. Oxford: Oxford University Press.

Lee, D. T. and C. K. Wong (1977). Worst-case analysis for region and partial region searches in multidimensional binary search trees and balanced quad trees. *Acta Informatica 9*(1), 23–29.

Lombard, K. and R. L. Church (1993). The gateway shortest path problem: Generating alternative routes for a corridor location problem. *Geographical Systems 1*(1), 25–45.

Mandelbrot, B. B. (1967). How long is the coast of Britain? Statistical self-similarity and fractional dimension. *Science 155*, 636–638.

Mandelbrot, B. B. (1977a). *The Fractal Geometry of Nature*. San Francisco: W. H. Freeman.

Mandelbrot, B. B. (1977b). *Fractals: Form, Chance, and Dimension*. San Francisco: W. H. Freeman.

Manolopoulos, Y., A. Nanopoulos, A. N. Papadopoulos, and Y. Theodoridis (2006). *R-Trees: Theory and Applications*. London: Springer.

Matheron, G. (1963). Principles of geostatistics. *Economic Geology 58*, 1246–1266.

McGarigal, K. and B. J. Marks (1995). FRAGSTATS: Spatial pattern analysis program for quantifying landscape structure. Technical Report PNW-GTR-351, U.S. Department of Agriculture, Forest Service, Pacific Northwest Research Station.

Meagher, D. (1980). Octree encoding: A new technique for the representation, manipulation and display of arbitrary 3-D objects by computer. Technical Report IPL-TR-80-111, Rensselaer Polytechnic Institute.

Meagher, D. (1982). Geometric modeling using octree encoding. *Computer Graphics and Image Processing 19*, 129–147.

Michalewicz, Z. and D. B. Fogel (2000). *How to Solve It: Modern Heuristics*. Berlin: Springer.

Miller, H. J. and S.-L. Shaw (2001). *Geographic Information Systems for Transportation: Principles and Applications*. New York: Oxford University Press.

Moran, P. A. P. (1950). Notes on continuous stochastic phenomena. *Biometrika 37*(1), 17–23.

O'Neill, R., J. Krummel, R. Gardner, G. Sugihara, B. Jackson, D. DeAngelis, B. Milne, M. Turner, B. Zygmunt, S. Christensen, V. Dale, and R. Graham (1988). Indices of landscape pattern. *Landscape Ecology 1*(3), 153–162.

O'Rourke, J. (1998). Point in polygon. In *Computational Geometry in C* (2nd edn), pp. 279–285. Cambridge: Cambridge University Press.

Peitgen, H.-O., H. Jürgens, and D. Saupe (1992). *Chaos and Fractals: New Frontiers of Science*. New York: Springer.

Peitgen, H.-O. and D. Saupe (1988). *The Science of Fractal Images*. New York: Springer.

Plastria, F. (2002). Continuous covering location problems. In Z. Drezner and H. W. Hamacher (eds), *Facility Location: Applications and Theory*, pp. 37–79. Berlin: Springer.

Press, W. H., S. A. Teukosky, W. T. Vetterling, and B. P. Flannery (2002). *Numerical Recipes in C++* (2nd edn). Cambridge: Cambridge University Press.

Resende, M. G. C. and R. F. Werneck (2003). On the implementation of a swap-based local search procedure for the p-median problem. In R. E. Ladner (ed.), *Proceedings of the Fifth Workshop on Algorithm Engineering and Experiments*, pp. 119–127. Philadelphia: SIAM.

Resende, M. G. C. and R. F. Werneck (2004). A hybrid heuristic for the p-median problem. *Journal of Heuristics 10*, 59–88.

Rey, S. J. and L. Anselin (2010). PySAL: A Python library of spatial analytical methods. In M. Fischer and A. Getis (eds), *Handbook of Applied Spatial Analysis: Software Tools, Methods and Applications*, pp. 175–193. Berlin: Springer.

Richardson, R. T. (1989). Area deformation on the Robinson projection. *American Cartographer 16*, 294–296.

Riitters, K. H., R. V. O'Neill, J. D. Wickham, and K. B. Jones (1996). A note on contagion indices for landscape analysis. *Landscape Ecology 11*(4), 197–202.

Ripley, B. D. (1981). *Spatial Statistics*. New York: John Wiley & Sons.

Robinson, A. H. (1974). A new map projection: Its development and characteristics. *International Yearbook of Cartography 14*, 145–155.

Rosing, K. E. (1997). An empirical investigation of the effectiveness of a vertex substitution heuristic. *Environment and Planning B: Planning and Design 24*(1), 59–67.

Rosing, K. E., E. L. Hillsman, and H. Rosing-Vogelaar (1979). A note comparing optimal and heuristic solutions to the p-median problem. *Geographical Analysis 11*(1), 86–89.

Rosing, K. E. and M. J. Hodgson (2002). Heuristic concentration for the p-median: An example demonstrating how and why it works. *Computers & Operations Research 29*, 1317–1330.

Samet, H. (1990a). *Applications of Spatial Data Structures: Computer Graphics, Image Processing, and GIS*. Reading, MA: Addison-Wesley.

Samet, H. (1990b). *The Design and Analysis of Spatial Data Structures*. Reading, MA: Addison-Wesley.

Samet, H. (2006). *Foundations of Multidimensional and Metric Data Structures*. San Francisco: Morgan Kaufmann.

Samet, H. and R. E. Webber (1985). Storing a collection of polygons using quadtrees. *ACM Transactions on Graphics 4*(3), 182–222.

Sellis, T., N. Roussopoulos, and C. Faloutsos (1987). The R+-tree: A dynamic index for multi-dimensional objects. Technical report, VLDB Endowments.

Shamos, M. and D. Hoey (1976). Geometric intersection problems. In *17th Annual Symposium on Foundations of Computer Science*, pp. 208–215. IEEE.

Sinnott, R. (1984). Virtues of the Haversine. *Sky and Telescope 68*, 159.

Snyder, J. P. (1987). Map projections – a working manual. Professional paper 1395, U.S. Geological Survey.

Snyder, J. P. (1990). The Robinson projection: A computation algorithm. *Cartography and Geographic Information Systems 17*(4), 301–305.

Sylvester, J. J. (1860). On Poncelet's approximate linear valuation of Surd forms. *Philosophical Magazine 20*, 203–222.

Teitz, M. B. and P. Bart (1968). Heuristic methods for estimating the generalized vertex median of a weighted graph. *Operations Research 16*, 955–961.

Tobler, W. R. (1970). A computer movie simulating urban growth in the Detroit region. *Economic Geography 46*, 234–240.

Webster, R. and M. A. Oliver (1990). *Statistical Methods in Soil and Land Resource Survey*. Oxford: Oxford University Press.

Welzl, E. (1991). Smallest enclosing disks (balls and ellipsoids). In H. Maurer (ed.), *New Results and New Trends in Computer Science*, pp. 359–370. Berlin: Springer.

Whitaker, R. (1983). A fast algorithm for the greedy interchange of large-scale clustering and median location problems. *INFOR 21*, 95–108.

Wilford, J. N. (1988). The impossible quest for the perfect map. *New York Times*, October 25. http://nyti.ms/1BcUqAt.

Wolpert, D. and W. Macready (1997). No free lunch theorems for optimization. *IEEE Transactions on Evolutionary Computation 1*(1), 67–82.

Xiao, N. (2006). An evolutionary algorithm for site search problems. *Geographical Analysis 38*(3), 227–247.

Xiao, N. (2008). A unified conceptual framework for geographical optimization using evolutionary algorithms. *Annals of the Association of American Geographers 98*(4), 795–817.

Xiao, N. (2012). A parallel cooperative hybridization approach to the *p*-median problem. *Environment and Planning B: Planning and Design 39*, 755–774.

Xiao, N., D. A. Bennett, and M. P. Armstrong (2002). Using evolutionary algorithms to generate alternatives for multiobjective site search problems. *Environment and Planning A 34*(4), 639–656.

Xiao, N., D. A. Bennett, and M. P. Armstrong (2007). Interactive evolutionary approaches to multiobjective spatial decision making: A synthetic review. *Computers, Environment and Urban Systems 30*, 232–252.

Xiao, N. and A. T. Murray (2015). Interpolation. In M. Monmonier (ed.), *History of Cartography, Volume 6: Cartography in the Twentieth Century*. Chicago: University of Chicago Press.

Young, C., D. Martin, and C. Skinner (2009). Geographically intelligent disclosure control for flexible aggregation of census data. *International Journal of Geographical Information Science 23*(4), 457–482.

Index

Note: Page numbers in **bold** indicate a more extensive treatment of the topic.

1-centre location problem **230–44**, 248, 274
   algorithms
      Chakraborty–Chaudhuri 233–4, 243
      Chrystal–Peirce 233, 234–6, 243
      Elzinga–Hearn 237–40
      testing and comparing 242–4
      Welzl 240–2
   data structures 231–3
2-D trees 77
adjacency matrix 193–4, 298–300
Albers projection 295, 297
algorithms, nature of 1–2
all pair shortest paths 222–5
arc distance 14
area, polygon 18–20
autocorrelation, spatial **191–9**
AVL tree 54

Bentley, J. L. 97, 110
Bentley–Ottmann algorithm **51–8**, 64
binary search 5, 72–3
binary tree
   computational cost 4–6
   and indexing 71–6
   searching 5, 72–3
   structure 4, 72
breadth-first traversal 214–16
brute-force approach 1, 49, 96
   range values 153
buckets 97

centroid 18
Chakraborty–Chaudhuri method 233–4, 243, 248
Chrystal–Peirce algorithm 233, 234–6, 243, 248

circular range query 84, 90–1, 107–8
classes, Python 288–90
clustering **199–202**
coding 6–7
   code and data lists 303–6
computational cost 2–6
   heuristic algorithms 272
   and running time notation 4
   solution space search 229–30
   and spatial indexing 96–7
contagion 202
   calculating 202–4
   testing 205–7
contours, interpolation of 141, 175
cost *see* computational cost
covariance 151, 158, 159, 163
covering problem 250
CPLEX 245
cross-validation 168–9

data structure 6
DCEL *see* doubly connected edge list
depth-first traversal 216–17
Dijkstra's algorithm 217–21, 225
distance
   between two points 13–15
   computational cost of calculation 3–4
   point to line 15–18
   Python code 282
distance matrices 222–4, 257
doubly connected edge list (DCEL) 58–60, 66, 113
   extended (XDCLE) 113–15

edges, network 211–12
Elzinga–Hearn algorithm 237–40, 248

Euclidean distance 12, 13
event queue module 52–3
even–odd algorithm 27–30
exact interpolation 161

factorial calculation 282–3
first law of geography 141, 175
first-in-first-out (FIFO) rule 214
Floyd–Warshall algorithm
    222–5, 226
format, Python 279
fractal interpolation 172–4, 175
functions, Python 282–4

gateway problem 221, 226
GDAL 291–2, 300–1
geographic coordinate system 294
geospatial data 291–2, 295–8
Google Maps 96
great circle distance 14
greedy algorithms 251–3
GUROBI 245

Haversine formula 14, 45
heuristic methods 230, 251, 273
    see also greedy algorithms; simulated annealing; vertex exchange
hybrid algorithms 273

IDW see inverse distance weighted
indentation, Python 280–1
indexing 71–6, 112
    code list 304
    time issues 96
interpolation
    code list 304–5
    general discussion of 139–41, 174–5
        exact interpolation 161
        methods compared
            165–70, 174
    specific approaches to see inverse distance weighted; kriging; midpoint displacement
intersection
    between two lines 22–3
    line segments 23–5, 49–57
    and point-in-polygon tests 27–8
inverse distance weighted (IDW) interpolation
    141–6, 175
    cross validation 145–6
    data preparation 143–5

inverse distance weighted (IDW) interpolation *cont.*
    power of distance value 142, 144–5
    use to interpolate a surface 165–70

k-D trees **77–97**
    point k-D trees 77–87
    point region (PR) k-D trees 87–93
    testing k-D trees 93–7
    time and computational issues 96–7
    use of buckets 97
K-function **185–91**, 208
k-means algorithm 199–202
kernel functions 206–7
Krige, D. 174
kriging **146–70**, 174–5
    concept of semivariance 147
        empirical computation 147–50
        modeling 150–6
    ordinary kriging 156–62, 167
    simple kriging 162–4, 167
    use to interpolate a surface 165–70

lag distance 150
lambda functions 283–4
landscape ecology 202
last-in-first-out (LIFO) rule 216
latitude 37–9
line segment intersection 23–5, **49–57**
linear search 2–4
lines
    on curved surface 14–15
    intersection of 22–5
    point distance from 15–18
    point position relative to 20–1
    see also line segment intersection
list comprehension, Python 280–1
location problems
    multiple location see p-median problem
    single location see 1-centre location problem
longitude 37–9
LP file format 245–7

Manhattan distance 14
map overlay **58–64**
map projections **33–45**
Matplotlib 41, 285–7
MBR 126–7, 135
Metropolis algorithm 261–2

midpoint displacement **170–5**
  diamond and square configurations 171
  fractal interpolation 172–4
minimal bounding rectangle (MBR) 126–7, 135
minimal enclosing circle *see* 1-centre location
minmax problem 249
modules, Python 282–3, 303
Mollweide projection **42–5**, 46
Monte Carlo simulations
  K-function 186–9
  Moran's I 195–7
  nearest neighbor distance 181–2
Moran's I **191–9**, 208
  computing 191–4
  Monte Carlo simulation 195–7
  testing 197–9
  variance 195

nearest neighbor queries
  k-D trees 85–7, 92–3
  point pattern analysis 178–85
  quadtrees 109–10
network analysis **211–25**
  code list 305
  network defined/described 211–14, 225–6
  network traversals 214
    breadth-first 214–16
    depth-first 216–17
  shortest path algorithms
    all pair 222–5
    single source 217–21
Neville's algorithm 37–9
nodes
  network 211–12
  tree 4–5, 72, 74–5
nugget and sill 150, 152, 153
NumPy 43, 101, 148, 160, 284–5

OGR **291–300**
  adjacency matrix 193–4, 298–300
  attributes 293
  geometry and coordinates 293–4
  projecting geospatial data 295–8
  projecting points 294–5
online maps 96
optimal prediction/interpolation 174
optimization *see* spatial optimization
ordinary kriging **156–62**, 167
  defined 158
  and stationarity hypothesis 157

orthogonal range query 82–3
overlay **58–64**

p-centre problem 249–50
p-median problem 244–8
  heuristic approaches to *see* greedy algorithms; simulated annealing; vertex exchange
pattern analysis *see* point pattern analysis
PM quadtrees *see* polygonal map quadtrees
point k-D tree **77–97**
  circular range query 84
  nearest neighbor query 85–7
  orthogonal range query 82–3
  testing 93–7
point pattern analysis **178–89**
  nearest neighbor analysis 178–85
    distance 179–81
    Monte Carlo simulation 181–2
    test function 181
  Ripley's K-function 185–91
    Monte Carlo simulation 186–9
point quadtrees **105–10**
  circular range search 107–8
  nearest neighbor query 109–10
point region (PR) k-D trees **87–97**
  circular range query 88
  nearest neighbor query 92–3
  testing 93–7
point-in-polygon operation **26–33**, 45–6
  even–odd algorithm 27–30
  winding number algorithm 30–3
points
  code list 303–4
  data structure of 11–13
  distance between 13–14
  plotting using Matplotlib 285–7
  point to line distance 15–18
  point-in-polygon tests 26–33, 45–6
  point.py class 12
  position with respect to a line 20–2
polygon
  area 18–20
  minimal bounding rectangle 126–7
  see also doubly connected edge list
polygon overlay algorithm 64–6
polygonal map quadtrees **112–26**, 136
  data issues 112–16
  PM1 quadtree 116–22

polygonal map quadtrees *cont.*
   PM2 quadtree 122–4
   PM3 quadtree 125–6
power of distance value 142, 144–5
PR trees *see* point region k-D tree; point region quadtree
PySAL 208–9, 301–2
Python **277–306**
   basic commands 277–81
   classes 288–90
      init method 288
      repr method 288
   functions 282–3
      lambda functions 283–4
      recursive functions 283
   list of code 303–6
   list comprehension 280–1
   Matplotlib 41, 285–7
   modules 282–3, 303
   NumPy 284–5
   packages/libraries 291–2
   simultaneous assignment 279

quadtrees **99–111**
   point quadtrees 105–10
   region quadtrees 99–105
   unbalanced/balanced 110–11
   see also polygonal map quadtrees
query function 79, 88
   query time 96
   see also range query

R-trees **126–36**
   minimal bounding rectangle 126–7, 135
   nodes and entries 128–30
   split and search functions 130, 133
randomness 181, 184–5, 189–90, 194
range 150, 153
range query functions
   circular range 84, 90–1, 107–8
   nearest neighbor 85–7, 92–3, 109–10
   orthogonal range 82–3
raster data 99–100, 287
   and GDAL 300–1
recursive functions 283
region quadtrees **99–105**
   query function 103–4
Ripley's K-function **185–91**, 208
RMSE 145, 168

Robinson projection **33–42**, 46–7
   testing with world map data 39–42
   transforming points to 36–9
root mean squared error (RMSE) 145, 168
root, tree 4–5, 72, 74
running time notation 4

scale, concept of 139–40
scatter plot 285–6
searches, computational cost of 2–6
semivariance **147–56**
   defined 147
   empirical computation 147–50
   modeling 150–6
   sill and nugget 150, 152, 153
semivariogram 149, 155
shapefile 292, 298
sill and nugget 150, 152, 153
simple kriging **162–4**, 167
simulated annealing **261–73**
   acceptance probability/optimal solutions 267–72
   p-median problem 262–7
single source shortest path 217–21
smallest enclosing circle *see* 1-centre location problem
software packages 7, 208–9
solution space 229–30
sorting, lambda function 284
spatial analysis, code list for 305
spatial autocorrelation **191–9**
spatial indexing *see* indexing
spatial optimization
   code list 305
   general issues 228–30, 244, 248
   specific approaches *see* 1-centre location problem; p-median problem
spatial patterns 177–8
spherical distance 14–15
stationarity hypothesis 157
subtrees 72
surface, interpolation of
   IDW and kriging 165–70
   midpoint displacement 170–4
sweep lines 50–3, 66–7
   Bentley–Ottmann algorithm 51–8, 64

Teizt–Bart algorithm *see* vertex exchange algorithm
time, computing *see* computational cost

Tobler's first law 141, 175
topography 141
   see also surface
trapezoids 18–19
traveling salesman problem 229
tree structures
   and computational cost 4–6, 96, 110
   and indexing 72–6
   two-dimensional trees 77, 78, 81
   unbalanced *versus* balanced 6–7, 97, 110–11
two-dimensional trees 77

vertex exchange algorithm **253–61**, 271–2, 273
   p-median problem 253–7
   testing problems in the OR-library 257–61

Welzl algorithm 240–2
winding number algorithm 30–3
world map data 40–2

XDCLE 113–15